Spacecraft Optical Navigation

DEEP SPACE COMMUNICATIONS AND NAVIGATION SERIES

Issued by the Deep Space Communications and Navigation Systems
Center of Excellence, Jet Propulsion Laboratory
California Institute of Technology

Jon Hamkins, Editor-in-Chief

Published Titles in this Series

Radiometric Tracking Techniques for Deep-Space Navigation
Catherine L. Thornton and James S. Border

Formulation for Observed and Computed Values of Deep Space Network Data Types for Navigation
Theodore D. Moyer

Bandwidth-Efficient Digital Modulation with Application to Deep-Space Communications
Marvin K. Simon

Large Antennas of the Deep Space Network
William A. Imbriale

Antenna Arraying Techniques in the Deep Space Network
David H. Rogstad, Alexander Mileant, and Timothy T. Pham

Radio Occultations Using Earth Satellites: A Wave Theory Treatment
William G. Melbourne

Deep Space Optical Communications
Hamid Hemmati, Editor

Spaceborne Antennas for Planetary Exploration
William A. Imbriale, Editor

Autonomous Software-Defined Radio Receivers for Deep Space Applications
Jon Hamkins and Marvin K. Simon, Editors

Low-Noise Systems in the Deep Space Network
Macgregor S. Reid, Editor

Coupled-Oscillator Based Active-Array Antennas
Ronald J. Pogorzelski and Apostolos Georgiadis

Low-Energy Lunar Trajectory Design
Jeffrey S. Parker and Rodney L. Anderson

Deep Space Communications
Jim Taylor, Editor

Radio Science Techniques for Deep Space Exploration
Sami W. Asmar

Spacecraft Optical Navigation
William M. Owen Jr.

Spacecraft Optical Navigation

William M. Owen Jr.
Jet Propulsion Laboratory
California Institute of Technology
California, USA

WILEY

Published by John Wiley & Sons, Inc., Hoboken, New Jersey.
Published simultaneously in Canada.

For general information on our other products and services or for technical support, please contact our Customer Care Department within the United States at (800) 762-2974, outside the United States at (317) 572-3993 or fax (317) 572-4002.

Wiley also publishes its books in a variety of electronic formats. Some content that appears in print may not be available in electronic formats. For more information about Wiley products, visit our web site at www.wiley.com.

Library of Congress Cataloging-in-Publication Data applied for:

Hardback ISBN: 9781119904434

Cover Design: Wiley
Cover Image: © Andrey Armyagov/Shutterstock

Set in 9.5/12.5pt STIXTwoText by Straive, Chennai, India

Contents

List of Figures

Preface

The book which you, my dear reader, are now holding in your hands represents not only the summation of the author's 40-year career at the Jet Propulsion Laboratory but also the accumulated wisdom, gained from experience, of many of my past and current colleagues. A brief autobiography may help provide some context for the rest of the material.

I was in high school when Mariners 6 and 7 flew past Mars and just out of college when the Vikings landed on it. I came to JPL in December 1979, after the Voyagers' encounters with Jupiter and before they flew past Saturn. I hired into what was then called the Optical Measurements Analysis Group, and they set me to work finding opportunities for optical navigation pictures for Galileo. At that time launch was scheduled for 1982, with a flyby of Mars, and I led a small effort to improve the positions of stars behind Mars before the flyby. (I still have some plates taken by the 48-inch Schmidt telescope at Palomar.) The launch slipped, and that work went for naught.

By 1985 I was finally supporting Voyager, not in operations but rather in star catalog development. Before the HIPPARCOS mission, the newest catalogs had been produced by the Yale University Observatory in the 1930s, and the uncertainties in the stars' proper motion had rendered their positions obsolete. The Voyager project, for each encounter, had a contract with Lick Observatory (by then managed by the University of California, Santa Cruz) to take glass plates, measure the stars on them, reduce the data, and produce an *ad hoc* catalog. Their state-of-the-art measuring engine was controlled by a PDP-8 minicomputer, but their setup required approximate plate coordinates for each image to be measured. A "survey machine," also connected to that computer, would project a small part of a plate onto a screen. There were crosshairs on the screen, a joystick to move the plate around, and a button (on a rack over to the side) labeled RECORD. I shall leave the rest of it to the reader's imagination. Suffice it to say that I surveyed eight plates for Uranus and enjoyed the experience so much that I repeated it for Neptune and for Galileo's flybys of Gaspra and Ida.

Changes to Voyager's approach sequence, coupled with a realization that the operations team could actually perform their tasks faster than originally planned, permitted us to schedule a few more late pictures. Those background stars lay outside the region we had catalogued, and I found myself writing to Southern Hemisphere observatories to request positions for these stars. The replies invariably began "Dear Dr. Owen," and not willing to make liars out of them, I went to grad school in fall 1986, studying astrometry under Heinrich Eichhorn, escaping just in time to support the Neptune flyby.

Since that time I have supported just about every mission that required optical navigation: Galileo (to a limited extent), Near Earth Asteroid Rendezvous (later renamed NEAR Shoemaker), Cassini, Deep Space 1, Stardust, Deep Impact, and New Horizons. I have also played a small role in some European missions, most notably Rosetta.

In the early 1990s I participated in some advanced studies to figure out how spacecraft navigation would work in the era of optical communication. Radio tracking data, so important for so many years, would be supplanted by a combination of LIDAR-style ranging and good old-fashioned optical astrometry, because a downlink laser would look just like a star in a telescope. We therefore started a groundbased astrometry campaign, first to characterize the instrument and camera, then to obtain data for the Galilean and Saturnian satellites, Mathilde and Eros, all in support of Galileo, Cassini, and NEAR. Our observations of Mathilde moved its orbit by something like eight standard deviations, a result that of course met with some resistance. David Dunham, now retired from APL but still active in the International Occultation Timing Association, was on NEAR's mission design team. He asked us for astrometry for one particular asteroid, (170) Maria, that was about to occult a star. We obliged, the predicted path of the occultation moved by 10 times its width, and someone in the revised path watched the event. This success gave the NEAR team the courage to accept our Mathilde astrometry, which also turned out to be correct.

The experience with Maria and Mathilde was the genesis of a collaboration with IOTA that would last twenty years.

All good things must end, so they say, and as I enter retirement it behooves me to thank all those who have helped me along the way:

- my supervisors, Ken Rourke, Tom Duxbury, and especially Steve Synnott;
- my graduate advisor, Heinrich Eichhorn, who taught this know-it-all even more than I thought possible;
- my many optical navigation colleagues over the years, particularly Ed Riedel, Bob Gaskell, and Nick Mastrodemos;
- my current management team, Ralph Roncoli and Tung-Han You, who have been holding my feet to the fire;

- David Berry, Bill Taber, and Dimitrios Gerasimatos, who provided financial support; and finally
- Richard Waldo Hambleton, my grandfather, who woke me up in the middle of one March night in 1960 to show me an eclipse of the moon.

Thank you, all of you.

Acknowledgement

The research described in this book was carried out at the Jet Propulsion Laboratory, California Institute of Technology, under contract to the National Aeronautics and Space Administration.

1

Introduction

1.1 Purpose

This book attempts to describe the discipline of spacecraft observation, as practiced at the Jet Propulsion Laboratory (JPL) since the late 1960s and more recently elsewhere. It is not intended to be a user's guide for JPL software, nor a mathematical models document describing the functionality of JPL software, nor a requirements document. Rather, it is meant to be more of a textbook treating all aspects of optical navigation: predicting the contents of pictures, analyzing the pictures to extract the observables, and introducing these observations into a solution for navigation parameters. Discussions of cameras and detectors are included as well, but only in enough depth that a navigation engineer can understand the properties of the hardware.

Chapter 2 contains the history of optical navigation, both at JPL and elsewhere. Chapter 3 discusses cameras, including their optics and detectors, from the point of view of an optical navigation engineer. Chapter 4 presents the mathematics required to predict where the image of some object or star will appear in a picture. Chapter 5, the longest and most diverse, discusses various methods of extracting observed image locations from the brightness levels in a received picture. Chapter 6 deals with the incorporation of optical data into an orbit determination solutions. Appendix treats the "overlapping plate method," which enables accurate camera calibration even though its original purpose was for the development of star catalogs. A glossary of terms and references for each chapter are also presented.

1.2 Definitions

Optical navigation is the use of pictures, taken by cameras aboard spacecraft, to help determine the trajectory of the spacecraft. A *camera* in this context means some instrument that projects incoming light onto a detector to form a *picture* of

Spacecraft Optical Navigation, First Edition. William M. Owen Jr.

the scene being imaged. The picture consists of a rectangular array of *pixels*, each of which contains a *digital number* or *DN* value that ideally is linearly proportional to the number of photons impinging upon a corresponding rectangular region of the detector. The picture may therefore contain *images* of stars, natural bodies, or even other spacecraft. *Landmarks* are specific features or points on the surface of a body, and images of landmarks are also used in navigation.

The fundamental coordinate system is the International Celestial Reference Frame or ICRF, currently defined at optical wavelengths by the coordinates of stars in the HIPPARCOS catalog (European Space Agency 1997). This frame's x-axis is closely aligned with the direction of the Earth's mean vernal equinox at epoch January 1, 2000 12:00:00 barycentric dynamical time (TDB), and its z-axis is closely aligned with the Earth's mean north pole at the same epoch. The origin is the barycenter of the Solar System. Star catalogs and ephemerides of Solar System bodies used in optical navigation operations are presumed to be referred to the ICRF as well.

1.3 Notation

Throughout this book, scalars are set in *italic* type, vectors in **boldface roman**, and matrices in uppercase sans-serif. Terms specific to astronomy, optics, or optical navigation appear in *slanted* type at their first occurrence.

1.4 Rotations

Rotations are to be thought of as rotating a *coordinate system*, rather than as a change in the orientation of (say) a vector. A *rotation matrix* is a 3×3 matrix that, when multiplying a vector, transforms its rectangular coordinates from one coordinate system to another. An "elementary" rotation by angle θ about axis i is denoted as $\mathsf{R}_i(\theta)$. Written out in full,

$$\mathsf{R}_1(\theta) = \begin{pmatrix} 1 & 0 & 0 \\ 0 & \cos\theta & \sin\theta \\ 0 & -\sin\theta & \cos\theta \end{pmatrix}, \tag{1.1a}$$

$$\mathsf{R}_2(\theta) = \begin{pmatrix} \cos\theta & 0 & -\sin\theta \\ 0 & 1 & 0 \\ \sin\theta & 0 & \cos\theta \end{pmatrix}, \tag{1.1b}$$

$$\mathsf{R}_3(\theta) = \begin{pmatrix} \cos\theta & \sin\theta & 0 \\ -\sin\theta & \cos\theta & 0 \\ 0 & 0 & 1 \end{pmatrix}. \tag{1.1c}$$

Figure 1.1 shows a typical right-handed Cartesian coordinate system, whose axes are denoted by x, y, and z. Figures 1.2–1.4 show the result after a rotation

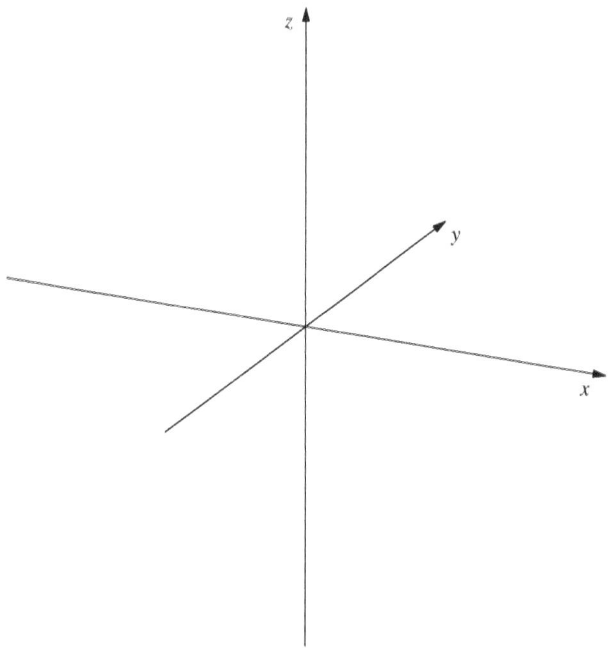

Figure 1.1 A right-handed coordinate system.

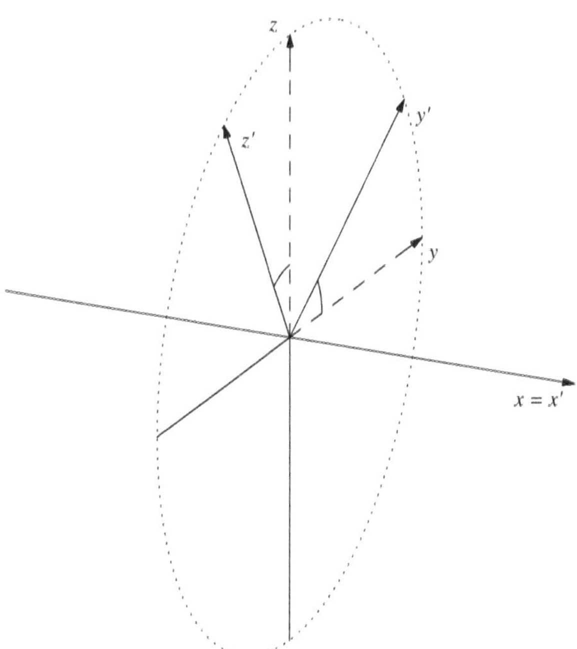

Figure 1.2 Rotation about the x-axis, by $+30°$.

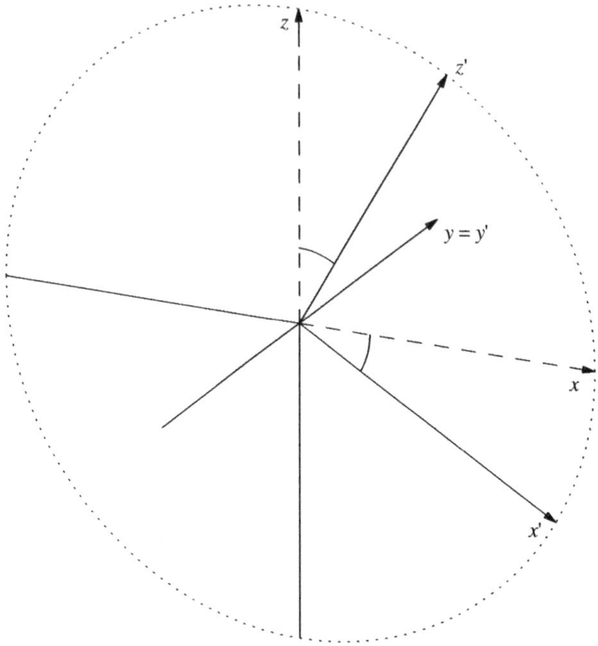

Figure 1.3 Rotation about the y-axis, by $+30°$.

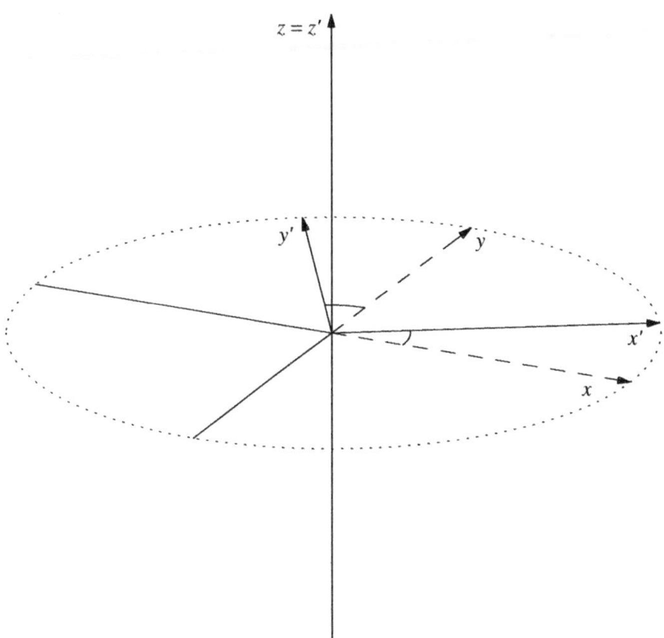

Figure 1.4 Rotation about the z-axis, by $+30°$.

of +30° about each of the three axes. The dotted circle in each of these figures represents the plane in which the rotation takes place—in other words, the plane which is perpendicular to the axis of rotation.

- A positive rotation about the x-axis moves the y-axis toward $+z$ and the z-axis toward $-y$.
- A positive rotation about the y-axis moves the z-axis toward $+x$ and the x-axis toward $-z$.
- A positive rotation about the z-axis moves the x-axis toward $+y$ and the y-axis toward $-x$.

If a vector $\mathbf{r}^A$ has rectangular coordinates that are expressed in some coordinate system A, the rotation matrix $\mathbf{T}_A^B$ will transform the coordinates into those of some other coordinate system B by

$$\mathbf{r}^B = \mathbf{T}_A^B \, \mathbf{r}^A. \tag{1.2}$$

The physical direction of the vector does not change under this operation. Its rectangular components, however, will in general change as a result of the change of coordinate system. The length of the vector is also unchanged, as the determinant of a rotation matrix is 1.

Rotation matrices may be strung together, one multiplying the next, to produce a new matrix which will also be a rotation. Three elementary rotations suffice to generate any arbitrary rotation. Matrix multiplication is associative but not commutative, and so the order in which the rotations are carried out is important. Figures 1.5–1.7 provide an example of a "3–1–3" sequence of rotations. The first rotation establishes the line of intersection between the initial and final x–y planes, the second gives the dihedral angle between these two planes (thereby establishing the final z-axis), and the third locates the final x- and y-axes. For example, if the final plane is the plane of an orbit, the three angles would be, in order, the longitude or right ascension[1] of the ascending node, the inclination, and the argument of pericenter.

Another rotation scheme in common use is 1–2–3, or some permutation thereof. This approach is particularly appropriate if the three rotation angles are very small, as is often the case for camera misalignment angles. If so, then the three rotations very nearly commute, and one can solve easily and simultaneously for the three angles.

---

1 These two terms are not synonymous. The *right ascension* of the ascending node is used when the fundamental plane is the earth's equator; the *longitude* of the ascending node is used not only when the fundamental plane is the ecliptic (as for planetary ephemeris theory) but also in general when the central body is not the earth or sun.

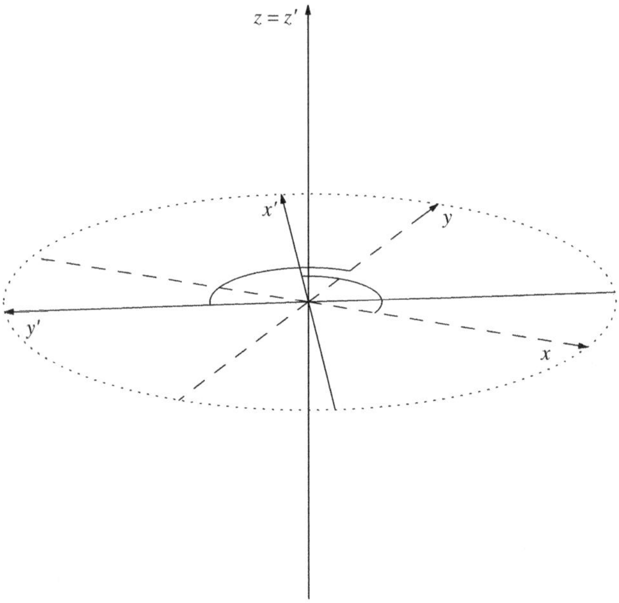

Figure 1.5 First rotation in a 3–1–3 sequence, by 120° about the *z*-axis. The dashed lines represent the original *x*- and *y*-axes.

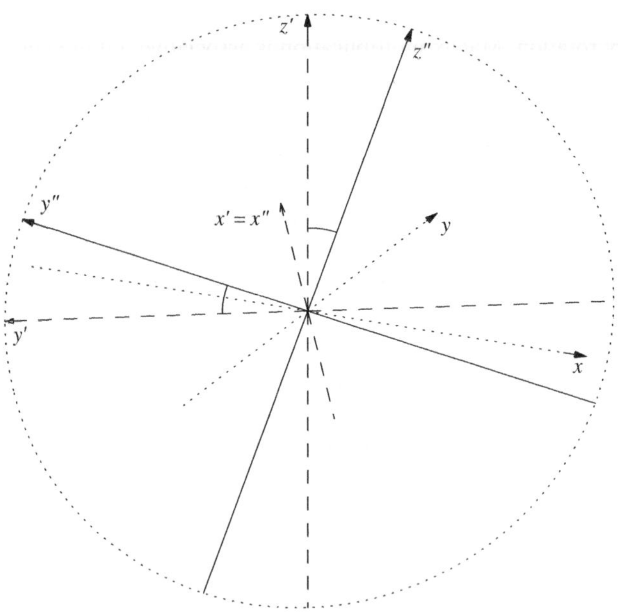

Figure 1.6 Second rotation in a 3–1–3 sequence by 20°, about the new *x*-axis. The original *x*- and *y*-axes now appear as dotted lines; the original *z*-axis and the *y*′-axis (i.e. after the first rotation) are dashed lines. Note that as the *x*-axis points into the page, a positive rotation appears to be clockwise from the reader's point of view.

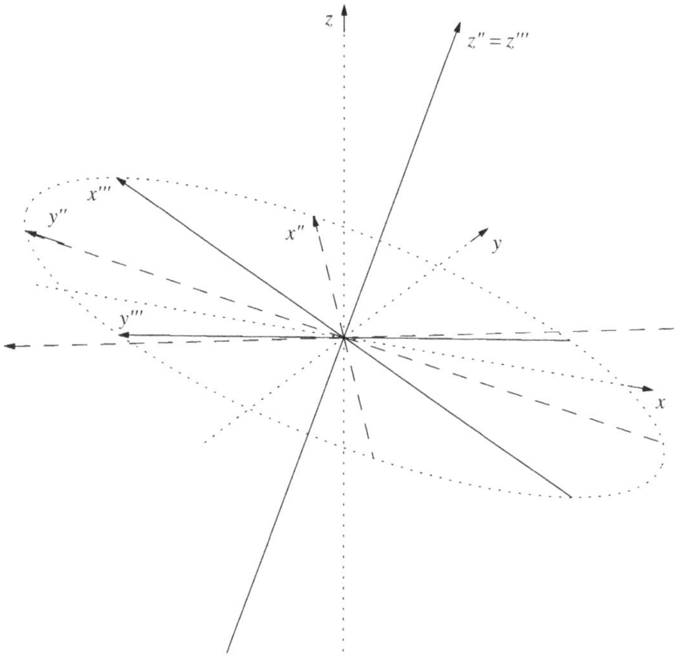

Figure 1.7 Third rotation in a 3–1–3 sequence, by 40° about the final z-axis.

1.5 Left-handed Coordinate Systems

Although the engineering world tends to use only right-handed coordinate systems, astronomers use three left-handed coordinate systems. Two are used primarily at observatories, but the third, treating planetary latitude and longitude, is definitely of concern to optical navigators. These three coordinate systems are:

1. *Topocentric azimuth and altitude.* The azimuth of a celestial object is measured from zero at due north, positive eastward. The altitude of a celestial object is its angular distance above the local horizon. (Some people incorrectly call this the "elevation," but for astronomers the term *elevation* is the observer's distance above sea level.) These two definitions imply that in the topocentric coordinate system, the x- and y-axes are horizontal, with $+x$ pointing north and $+y$ pointing east. The $+z$-axis of course must point to the zenith. This convention produces a left-handed system.
2. *Topocentric hour angle and declination.* The hour angle measures how far an object lies east or west of the meridian. As the earth rotates eastward, stars and other celestial objects appear to move westward. (It's not only the sun that rises

in the east and sets in the west.) As angles traditionally increase with time, the hour angle must be negative for objects to the east of the meridian and positive for those to the west. The corresponding coordinate system must therefore have the $+x$-axis pointing to the intersection of the celestial equator and the meridian, the y-axis pointing to the west point on the horizon (azimuth 270°) and the $+z$-axis pointing to the celestial north pole. This is therefore a left-handed system.

3. *Latitude and longitude for planets in direct rotation.* The International Astronomical Union has decreed that "the north pole is that pole of rotation that lies on the north side of the invariable plane of the Solar System" (Archinal *et al.* 2010) and further that the longitude of the "central meridian" of a planet should *increase with time*, as seen by an external observer. As these planets rotate from west to east, like the Earth, the longitude must accordingly increase to the west. Therefore the $+z$-axis points to the planet's north pole, the $+x$-axis lies at the intersection of the equator and prime meridian (however defined), and the $+y$-axis must point to 90° *west* longitude. Again, this is a left-handed system. Note, however, that Venus and Uranus, in retrograde rotation, will have a right-handed system. Note also that dwarf planets and minor planets are not covered by this definition; rather, their coordinate systems are always right-handed, even if (as for Pluto) their rotational angular momentum vector points south of the invariable plane.

2

History

Much of the material in this chapter was originally presented in Owen *et al.* (2008). It has been updated for this book.

Optical navigation got its start at Jet Propulsion Laboratory (JPL) as an experiment on the Mariner 6 and 7 missions to Mars in 1969 (Duxbury and Breckenridge 1970, Duxbury 1970) and again on Mariner 9 in 1971 (Breckenridge and Acton 1972). The justification for opnav was to ensure quality navigation results at the outer planets: as radio tracking data are geocentric, a radio-only orbit determination solution will tend to become less accurate with increasing distance from the earth.

Opnav was used operationally for both Viking orbiters at Mars, but it came into its own with the Voyager missions to the outer planets. Pictures of the Galilean satellites of Jupiter, of Titan and the smaller icy satellites of Saturn, of the five classical Uranian satellites, and of Triton, Nereid, and Proteus (S/1989 N 1) at Neptune helped immensely to shrink the size of the B plane error ellipse.[1] Optical navigation engineers were also responsible for several of Voyager's discoveries. The serendipitous discovery of volcanic plumes on Io is best known, but optical navigators also found new satellites at all four outer planets and determined their orbits.

Both hardware and software advances have occurred since the 1970s. Vidicon television cameras have given way to charge-coupled detectors (CCDs) and complementary metal-oxide semiconductors (CMOS) detectors. Ground data processing has moved from mainframes to minicomputers and now to workstations. We once used special frame buffers and monitors for display; now, our workstations bring up pictures inside an X window. The software has progressed from a mixture of Fortran 66 and assembly language to Fortran 77 and C, and much of it has been rewritten in C++ and Python.

1 The "B plane" (Kizner 1961), defined for a spacecraft which is about to make a hyperbolic flyby of some object, is perpendicular to the incoming asymptote of the hyperbola. Navigation engineers usually control the spacecraft's trajectory in terms of where and when this asymptote pierces the B plane.

Spacecraft Optical Navigation, First Edition. William M. Owen Jr.

Perhaps the most promising advance in optical navigation technology is the migration from ground processing to onboard processing. JPL's onboard autonomous navigation system was demonstrated on Deep Space 1 and used on Stardust. Autonav's greatest success to date was on Deep Impact, where it ran on both the Impactor and Flyby spacecraft and guided the Impactor onto a collision course with the nucleus of comet 9P/Tempel 1 while the Flyby spacecraft was taking pictures of the event.

Opnav figures to play a prominent role in many future JPL missions: to small bodies, to Uranus and Neptune, and even to support precision landings on the moon or Mars. Whether the processing is done on the ground or onboard, whether the imager is like the dedicated Optical Navigation Camera flown on Mars Reconnaissance Orbiter or a science instrument, optical navigation data will continue to enable the kind of precision navigation which is required for mission success.

Optical navigation as practiced at JPL usually requires knowledge of several disciplines and interactions with both the flight team and the science team at various phases of a mission. Opnav analysts must know enough about the optical and physical characteristics of the onboard cameras to be able to simulate images and command pictures correctly. We must participate in orbit determination (OD) studies to find out how much optical data is necessary to meet navigation accuracy requirements; then we must negotiate for observing time, plan the pictures to be taken, develop the sequence products, and verify their correctness through the uplink process. After the pictures have been obtained, we must process them, extracting the observed locations of the images, and pass the results along to the rest of the navigation team. The optical navigation group must therefore have people who know about optics, some facets of astronomy, surface modeling, spacecraft commanding, image processing, and spacecraft navigation.

2.1 The Early Years: Mariner and Viking

The first so-called "optical navigation experiment" was carried out in 1969 on Mariners 6 and 7 to Mars. Tom Duxbury and the late Bill Breckenridge used the "far-encounter planet sensor" to take pictures (50 for Mariner 6, 93 for Mariner 7) of Mars in the last two or three days before each spacecraft's flyby (Duxbury and Breckenridge 1970). Figure 2.1 shows one such image. They measured the location of Mars in each picture by putting a clear plastic overlay on top of a hard copy of the picture, matching a circle on the overlay to Mars, and reading off the coordinates. (The paper makes a point of saying that each picture was measured by two "observers" and the results averaged.) The Mars image centers thus obtained were transferred to punched cards.

There were no stars in the pictures; the spacecraft attitude came from real-time telemetry. Careful calibrations before launch and during cruise had given the

Figure 2.1 Mariner 1969 picture of Mars. Source: NASA.

orientation of the camera relative to the spacecraft. The calibrations also revealed distortions in the camera, which were used to correct the observed image locations.

The image location and camera attitude were then transformed into the observed inertial direction to Mars. The right ascension and declination were the optical observables (not the image coordinates as is now the case). These angles were fed into a navigation filter which read in the current radio OD solution (from magnetic tape) and updated it.

Real-time opnav operations were successful. The pictures were received and processed within the allotted time. The resulting solutions were generally within one sigma of the radio-only solutions, and the major axis of the B-plane error ellipse shrunk by half.

After encounter, Duxbury devised a new technique (Duxbury 1970) for determining the center of Mars. Each line of the picture was scanned to find the lit limb, defined as the first of three successive pixels brighter than a threshold. As the limb of an ellipsoidal body projects into an ellipse in the focal plane, they fit an ellipse to the limb points, subject to *a priori* constraints on its shape from the known shape of Mars and the viewing geometry. The center of the ellipse was identified with the center of the planet. There were trends in the limb residuals indicating systematic effects, particularly near the poles, but the resulting data were good to about 1 pixel east–west and 0.3 pixel north–south. The worse performance in the horizontal direction was attributed to the fact that only the lit limb was used.

2.1.1 Mariner 9

Opnav was demonstrated again in the next mission, Mariner 9 in 1971. Tom Duxbury and Chuck Acton oversaw the creation of two sets of programs: the Optical Navigation Image Processing System (ONIPS), for picture display and image centerfinding, and the Optical Navigation Program set (ONP), for scene prediction, calculation of residuals and partial derivatives with respect to parameters of interest, and filtering. They planned 18 pictures of Deimos and 3 of Phobos, to be taken between 66 and 9 h before Mars orbit insertion. Deimos, farther from Mars, was the preferred target, and two of the Phobos pictures were timed to capture the satellite in transit across Mars. Several sets of calibration pictures, obtained during cruise, served to characterize the performance of the spacecraft attitude system, various camera misalignment angles, electromagnetic distortion in the camera's vidicon scan pattern, and the overall sensitivity of the camera.

Each of the 21 opnav pictures was analyzed within an hour of its receipt. Stars as faint as magnitude 8.9 were detected in 6-s exposures. The rms star residual, in pictures containing more than two stars, was 0.4 pixel. Results from the first 14 pictures were passed to the navigation team, which thereby obtained a more accurate OD at Mars than any previous mission. Improvements to the satellite ephemerides also enabled close-up imagery of Phobos and Deimos during the orbital phase of the mission.

The real benefit of Mariner 9, though, was the development of the opnav camera models, processing techniques, and team procedures. Much of what we do today traces its roots to 1971. ONIPS and ONP are still used, and even some of the program names are unchanged though the code has been rewritten several times.

Duxbury and Acton received NASA's Exceptional Scientific Achievement Medal for this work. They were also accorded the Institute of Navigation's Samuel M. Burka award for their paper (Duxbury and Acton 1972), along with the princely sum of $175 each – but they had to pay their own way to the ceremony (Figure 2.2).

2.1.2 Viking

The two Viking missions (1976) used optical navigation in operations not only on approach but also in orbit (Jerath 1978). The approach OD and the insertion

Figure 2.2 Tom Duxbury and Chuck Acton receiving the Samuel M. Burka award. Source: NASA.

maneuver were so accurate that the Viking 1 orbiter overflew the proposed landing site on its first orbit, rendering unnecessary two weeks of contingency operations. Dozens of opnav pictures taken by the orbiters enabled close encounters with both Martian satellites; a 20 km flyby of Deimos was in error by less than 2 km. Opnav was also useful in the orbiters' extended mission: as the amount of radio tracking data decreased, radio-only solutions became less sensitive to the node of the spacecraft's orbit on the plane of the sky. Pictures of the satellites provided a complementary data type which served well to reduce this uncertainty.

Navine Jerath and Ken Rourke formed the foundation of the opnav team for Viking.

2.2 Coming of Age: Voyager

The Voyager mission to Jupiter (1979), Campbell *et al.* (1980), Saturn (1980, 1981), and Campbell *et al.* (1982) was the first JPL project for which optical navigation was needed in order for the navigation accuracy to satisfy mission requirements: the greater distances in the outer Solar System mean that radiometric data alone would not suffice. Opnav was even more critical for Voyager 2's extended missions to Uranus (1986), Synnott *et al.* (1986), Neptune (1989), and Riedel *et al.* (1990). At Uranus, radio-only OD would have produced B-plane uncertainties of 2000–5000 km; the addition of optical data brought the delivery error under 100 km (Van Allen, Cesarone and Gray 1982). Similar results at Neptune enabled Sun and Earth occultations by Triton.

The opnav group obtained a pair of Modcomp IV minicomputers for image processing and rewrote the ONIPS software. New capabilities included the use of *radial* limb scanning to determine the centers of resolved bodies and "box filtering" (finding the maximum of a convolution of the actual image with a pixellated point spread) for stars and unresolved bodies. The ONP was also rewritten, in SFTRAN, a local flavor of "structured Fortran" (actually a precompiler). The opnav team at JPL even obtained their own star catalogs from photographic plates taken and measured at Lick Observatory.

The opnav team obtained the opnav pictures in negotiations with the various science teams during sequence development. Once the approximate shutter times were known, the opnav team would determine the camera pointing and pass the angles to the sequence team in the form of a paper memorandum. The sequencers would then create the commands to point the camera, set the various camera parameters, change the telemetry mode if necessary, and take the exposure. The opnav team would verify the timing and pointing in the resulting sequence products.

The Voyager scientists were generally willing to accommodate opnav requests at specific times when the stars were favorable. At Jupiter and Saturn, the star background was relatively sparse, and at Jupiter ("only" 5 AU from the Sun) we usually had to overexpose the satellites in order to get adequate signal on the reference stars. On approach to Uranus, however, the planet and its satellites appeared in front of the Milky Way in Sagittarius, the exposure times were longer, and almost any time would work. By the time we got to Neptune, the exposure times for science images were 15 s long, and stars usually appeared in them too, and so we used science images as well as opnavs.

The limb and terminator of the "gas giant" planets are difficult to measure accurately: the thick atmosphere does not produce a sharp boundary as a solid body does. We therefore used only satellites as targets.

At various times, Steve Synnott, Chuck Acton, Linda Morabito (Kelly), Homa Taraji, Adriana Ocampo, Ed Riedel, Joe Donegan, Juli Stuve, Bill Owen, and Robin Vaughan worked on Voyager opnav. The story of Morabito's serendipitous discovery of the volcanoes on Io (Morabito *et al.* 1979) is well known; what is less well known is that the discovery was made only because the satellite was highly overexposed.

Synnott (1980, 1981) discovered two new satellites in the Voyager 1 images at Jupiter (S/1979 J 2 and J 3, now named Thebe and Metis, respectively). The opnav software was quickly enhanced to recognize new bodies and to solve for their orbital elements, capabilities which persist to this day. Synnott would go on to find more "rocks" at Saturn, Uranus, and Neptune. Of these, S/1989 N 1 (Proteus) was most important from an engineering point of view. Neptune had only two known satellites before Voyager's arrival: massive Triton in a six-day retrograde orbit and small, distant Nereid. Covariance studies indicated that navigation performance would be greatly enhanced if a third satellite could be discovered and tracked for a month before encounter. Proteus turned up right on schedule, we determined its orbit, and the nav team used the data.

High-resolution images of many of these small satellites exist only because the opnav group could determine their orbits well enough and quickly enough to allow Voyager to take close-up images of them. After each encounter, we determined their orbits, using reconstruction trajectories and all available imaging, and published the orbits.

2.3 Innovation and Workarounds: Galileo

Galileo was launched in 1989, encountered two asteroids on its way to Jupiter, and orbited that planet from December 1995 to September 2003. Opnav for Galileo was originally designed to work much as it did for Voyager. Galileo's camera used a CCD instead of a vidicon as its detector, but the pictures were still 800 × 800 pixels with 8 bits per pixel. The same ONIPS software that worked so well for Voyager could be reused with some modification, so Riedel and Donegan rewrote it in SFTRAN. The ONP, however, was rewritten in HAL/S (the language used by the Shuttle's flight system) and ported to one of the two Modcomps. Frequent opnav images were to be taken on approach to the asteroids (951) Gaspra and (243) Ida, on approach to Jupiter, in orbit, and (in one launch scenario) on approach to Mars before a gravity assist. Special picture planning software, written by Bob Davis, scheduled pictures at times which would be best for determining not only Galileo's trajectory but also the orbits of the satellites, while ensuring a reasonable distribution of observations around the orbit and satisfying various mission constraints. The actual sequencing was still done by a sequence team, but the opnav inputs were now in electronic form, and the sequence products likewise were available electronically.

The Challenger explosion in January 1986 forced yet another launch delay, and by the time Galileo finally lifted off in 1989, the Modcomps were dead, HAL/S was moribund (at least at JPL), and the software, the ONP in Fortran 77 and ONIPS still in SFTRAN, was running on VAX computers.

The cruelest blow of all, though, was the failure of Galileo's high-gain antenna to open. The loss of the high gain antenna (HGA) meant that all communication had to use one of the low-gain antennas, at a much lower data rate. It was possible to downlink only a fraction of the desired opnav data – yet the navigation requirements did not change significantly. The only way out was to find ways of compressing the data, throwing away as much sky as possible in the pictures, retaining only the important parts. Ed Riedel, Robin Vaughan, and Mike Wang developed two workarounds.

The asteroid approach pictures were still returned as full frames (Vaughan *et al.* 1992), but instead of the usual short exposure the camera was commanded to take a very long exposure (Figure 2.3). While the shutter was open, the spacecraft was commanded to change its attitude in a succession of tiny slews. We expected that the target asteroid and the reference stars would show up as a set of bright spots with dimmer trails between them; we would perform centerfinding on each bright spot separately, in effect obtaining N independent measurements within

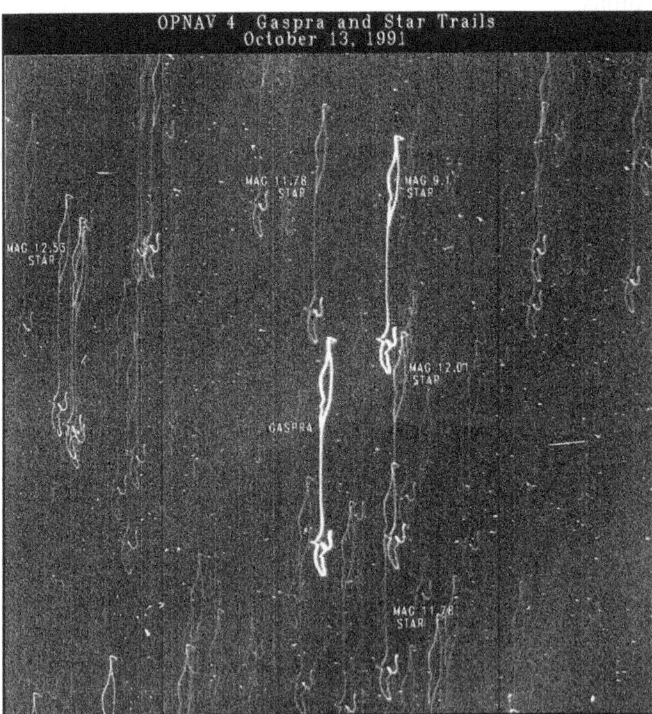

Figure 2.3 Galileo "single-frame mosaic" picture of Gaspra. Source: JPL/NASA.

one picture. In reality, the camera was always in motion to some degree, the images were trailed, but the trails had recognizable fiducial points which served quite well as independent measurements; in one case we could get 8 of them. (The centerfinding software needed an upgrade too: instead of looking for the actual image center, it had to locate the same fiducial point in each image using cross-correlation.) With this "single frame mosaic" technique, seven frames of Gaspra and three of Ida were multiplied into 17 and 19 independent pictures.

The second technique (Bhaskaran, Riedel and Synnott 1995) used Galileo's flight computers to do real-time image processing as the picture was being read off of the CCD. This technique required a massive programming of the camera's flight software as well as Galileo's command and data handling. The processing happened in several stages:

1. We would uplink to the spacecraft the expected limb locations of the target satellite, the expected data number (DN) patterns for horizontal scans across the limb, offsets from the center of the target to the reference stars, and other auxiliary data.
2. Each line of the picture was examined as it came off the CCD. If it contained several consecutive pixels whose DN values matched the predicted limb scan, these pixels were taken to be the top of the satellite image. Otherwise the line was discarded. Great pains were taken to ensure that the software would not be fooled by noise in the image.
3. Those pixels, with their surroundings, were kept; the remainder of the line was discarded.
4. Strips ("jail bars") surrounding the limb and terminator were retained from many (but not all) of the subsequent lines, until the bottom of the satellite image was reached.
5. The system computed the expected coordinates of the star images, based on the observed limb and the uplinked offsets. Because this computation could not be performed until the top of the satellite was detected, the stars had to appear below or to the side of the satellite.
6. Square arrays of pixels centered on the expected star images were retained as the appropriate lines were read out.

In this way Galileo was able to reduce greatly the number of pixels transmitted to the ground. While the number of pictures was significantly less than the original plan called for, we were able to salvage opnav operations and thereby meet the accuracy requirements for the updated satellite ephemerides. Figure 2.4 shows one such picture. The star – only one star! – appears near the center of its snip box, while the other bright spots are trails left by high-energy particles in Jupiter's environment.

Figure 2.4 A Galileo opnav picture, showing only those pixels which were transmitted to the ground. Source: NASA.

After the first four orbits of Jupiter, the satellite ephemerides and Galileo's trajectory were known well enough through radio tracking. The taking of opnav pictures ceased for the rest of the mission.

2.4 Landmarks: NEAR Shoemaker

JPL provided navigation services for the Applied Physics Laboratory's Near Earth Asteroid Rendezvous mission (later renamed "NEAR Shoemaker") to (243) Eros (Owen *et al.* 2001). Optical navigation was obviously going to be a critical component of the navigation system, as the weak gravity signal from Eros would not permit reliable radio-only OD. However, the standard opnav technique of locating the center of the target with respect to background stars would not work well. The high surface brightness of Eros meant that we'd have to overexpose it in order to bring out the stars, even more than Voyager did at Jupiter. The irregular shape of the asteroid would make it difficult to determine its center; in low orbit, the camera would see only a small part of the surface anyway. The solution was to use asteroid a year later. In hindsight, we could have done an adequate job in with the tools then available, but the year's delay proved to be a blessing. The n we had when we finally reached orbit in January 2000 was much better d to the task.

Maturity: Cassini

Cassini spacecraft, in orbit about Saturn from July 2004 to September used the traditional opnav technique of imaging satellites against a star ground (Gillam *et al.* 2007). Cassini's two cameras had CCD chips 1024 s square; the narrow-angle camera had a scale of 6 μrad or $1''\!.2$ per pixel. cameras produced 12-bit images. The imaging system's onboard computer a lossless compression algorithm, which reduced the data volume for opnav res by typically 75–80%. All these features were improvements over earlier nissions.

nav planning (at various times, Robin Vaughan, Claude Hildebrand, Bill n, Vijay Alwar, Shawna Santos, and Steve Gillam) proceeded along much the lines as for Voyager. Cassini, unlike Voyager or Galileo, had no scan plat- . The entire spacecraft had to turn in order to point the desired instrument ntenna) in the correct direction, all the while satisfying flight rules regarding ttitude and rates. As a result, no remote sensing observations could occur g downlinks, and the spacecraft was out of contact with the Earth during vations. (The NEAR spacecraft operated under similar constraints.) With satellites to choose from, there was almost always at least one with a good ield. In our negotiations for observing time, we usually restricted ourselves to s adjacent to downlinks, in order to maximize the contiguous time available cience.

quence development occurred in a distributed environment. Each science was responsible for submitting its own sequence files, and the spacecraft then merged these files into one integrated product. In order for this scheme ork smoothly, each observation had to begin and end at a specific "waypoint" de. Opnav, which submitted our files along with everyone else, had special ission to begin and end at two different attitudes (usually one was the link attitude and the other was the waypoint used for science). Our close ing relationship with the rest of the flight team let this practice work well, we thereby freed up quite a bit of turn time. Dennis Bluhm, Joan Pojman, Kathleen Kelleher created most of the opnav sequence products, but Drew han, Owen, and Gillam did their share too.

nav pictures began in earnest in February 2004, about four months before insertion. Opnavs of Phoebe proved essential (as expected) for Cassini's flyby at distant satellite in June. An intense observing campaign, lasting through 2005, improved the ephemerides of all the satellites. With the satellite orbits

identifiable surface features ("landmarks") and rely upon
craft attitude for camera orientation. NEAR's camera had p
(162 μrad on the longer side), and the attitude control sy
enough that the telemetered attitude agreed with the pictu

A crude landmark capability had existed in JPL's opnav s
but unused. A landmark was simply the tip of some vector c
of the body: with no size or shape, it was plotted in pictur
This capability was clearly inadequate, as we expected to be
cratered surface and needed some way of distinguishing or

The solution was to expand the definition of a landmark
radius and orientation, constant in body-fixed coordinates;
was the control point corresponding to the original definiti
solution proved to be more than adequate for the task (Fig

Measuring the centers of craters was another matter. We
surface modeling software. In 1998 we didn't. Instead, we
ONIPS picture display tool, to provide computer-aided feat
user would left-drag the cursor around the rim of a crater;
the screen pixels traversed by the cursor, fit an ellipse to th
it. If the results looked OK, the user would accept the me
would write it to a file for further processing. This was a t
got good at it, and our measurements of small, sharp crat
als of about a pixel. This was quite good enough to meet
one year's time, we obtained 131,337 observations of 1590
17,601 pictures.

Much of the software development for NEAR occurred
maneuver when NEAR first approached Eros in Decemb

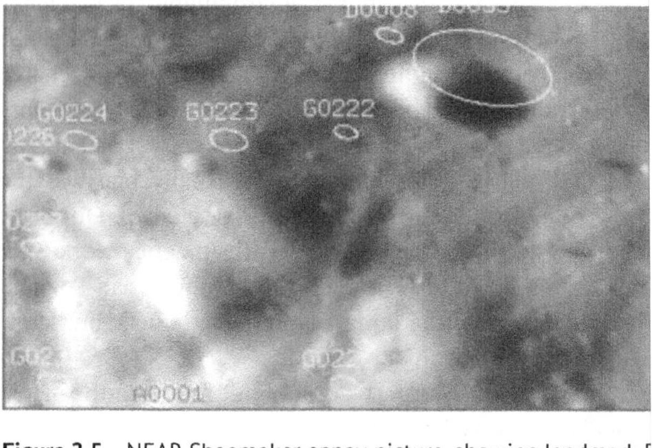

Figure 2.5 NEAR Shoemaker opnav picture, showing landmark [

to th
1998
syste
suite

2.5

The
2017
back
pixe
The
had
pict
JPL
O
Owe
sam
form
(or
the
duri
obse
nine
star
time
for s
Se
tean
tean
to w
atti
per
dow
wor
and
and
Vau
C
orb
of t
mic

well known and their mean motion errors brought down to an acceptable level, the use of opnav declined. Unlike Galileo, Cassini needed opnav support even into its extended mission. Only Titan was massive enough to show up in the doppler, and our frequent flybys of the icy satellites (in particular, Enceladus) force us to keep the satellite ephemerides in check. Image processing was done by Owen, Alwar, Mike Wang, Gillam, John Costello, Drew Vaughan, and Kevin Criddle.

Opnav operations for Cassini used the same network of Linux machines as the rest of the navigation team. All aspects of opnav, from picture planning to data reduction, happened on the same computer. File deliveries used the same large disk space as the rest of the team.

We had a few difficulties along the way. Titan, because of its thick atmosphere, proved to be a difficult target. Our observed centers had much larger random errors than we'd expected (but even Voyager had to deweight their Titan data), and there were unmistakable signs of phase-dependent bias in the sunward direction. Hyperion is in a chaotic rotation state due to its near resonance with Titan. We couldn't predict its orientation, and even after we allowed the centerfinding routine to step through all possible orientations, the results were sometimes poor. Iapetus is nearly white on the trailing hemisphere and nearly black ("Cassini Regio") on the other. These three satellites were such a nuisance that we stopped using Titan in 2006, and the other two were imaged in the extended missions only as a last resort.

We have also had our share of success stories. We used Cassini's cameras to image the Huygens probe after its release on December 24, 2004. Those pictures helped the nav team determine the direction and relative velocity of the two spacecraft and thus helped determine the subsequent trajectory of Huygens, which landed on Titan to much acclaim 22 days later.

2.6 Autonomy: Deep Space 1, Stardust, Deep Impact

Steve Synnott, Shyam Bhaskaran, Ed Riedel, and Bob Werner (among others) developed onboard autonomous navigation to handle encounters for which ground processing could not give adequate trajectory determination and control. Consider the situation of a spacecraft flying past a small body. Over a sufficiently short time, the relative trajectory is almost rectilinear, as the only significant perturbation is differential solar gravity. The target initially appears very near the velocity apex ("straight ahead" as seen from the spacecraft), and distant opnavs are sensitive only to the B-plane components, not to the arrival time. The error ellipsoid of the target, relative to the spacecraft, shrinks to a pencil whose long axis is along the velocity vector; the length is usually governed by the ground-based ephemeris error, the width by the angular resolution of the camera. As the spacecraft draws closer, the perspective begins to change, and opnav begins to "see" the downtrack direction in the pictures. At the moment of closest approach the target is perpendicular to the relative velocity. and the pencil is seen side-on. The highest-resolution imaging must therefore include many pictures of dark sky to ensure getting pictures of the target. If the opnav system and OD system can be put on the spacecraft, the round-trip light time to the ground is no longer a concern, and updates can be made almost continuously.

2.6.1 Deep Space 1

Deep Space 1, launched in 1998, was dedicated to flight-testing new technologies, one of which was such an onboard autonomous navigation system (Bhaskaran *et al.* 1998, 2000). The Autonav flight system, under the control of an executive, was in two-way communication with the spacecraft's attitude control system and camera. At predetermined times, it would command the ACS to point the camera at a desired target, command the camera to take a picture, and analyze the resulting picture, using the realized ACS attitude as input to the image analysis. Autonav also contained an OD filter, and it could command the ion propulsion system (another new technology) to control the trajectory.

During cruise, Autonav would take weekly pictures of several "beacon asteroids," brighter main-belt asteroids whose ephemerides were well known thanks in part to an intensive ground-based observing campaign at JPL's Table Mountain Observatory (Owen, Synnott and Null 1998). In effect these observations would provide an instantaneous position fix in three dimensions, and a sequence of observations spanning several months were sufficient to determine the spacecraft's trajectory to an accuracy of better than 1000 km and 0.4 m/s compared to ground OD solutions (Rayman *et al.* 2000).

On approach to the target, asteroid (9969) Braille, Autonav used a different scheme altogether. Initial images, taken only a day before closest approach, yielded the spacecraft's position in the B-plane, and we intended for Autonav to calculate and execute a maneuver at three hours before closest approach to control the aimpoint to an accuracy of 3 km (1σ). We also wanted Autonav to continue to execute after the maneuver, obtaining increasingly accurate estimates of the time of flight in order to control the timing of the science images. A failure in the flight version of Autonav kept it from doing the maneuver calculation (the ground version worked fine, and its results were used), and Braille's lower-than-expected albedo made it impossible for Autonav to detect the asteroid.

Deep Space 1 went on to an extended mission, a flyby of comet 19P/Borrelly. We fixed the problems from the Braille encounter, and this time Autonav performed flawlessly (Bhaskaran *et al.* 2002).

2.6.2 Stardust

The Stardust spacecraft was launched in 1999 and flew past asteroid (5535) Annefrank in 2002 on its way to an encounter with comet 81P/Wild 2 in 2004. Autonav for this mission was quite different than for DS1. It had no maneuvering capability: the encounter distance was large enough, and the delivery requirements loose enough, that the final maneuver could be computed on the ground. Rather, Autonav was used only to track the target during the time around closest approach (Bhaskaran, Riedel and Synnott 1997).

Autonav was initiated 20 min before encounter, using initial conditions computed by the navigation team. It commanded images (as a side note, the camera looked through a periscope, and the mirror orientation had to be commanded as well), reduced them, and fed the resulting image locations into a filter which estimated only the spacecraft position and attitude. The spacecraft relative velocity was assumed known.

Here too Autonav worked well (Bhaskaran *et al.* 2004), and Stardust captured the desired images of the cometary nucleus.

2.6.3 Deep Impact

The most demanding use of Autonav to date was the Deep Impact mission (2005) to send an Impactor spacecraft into the nucleus of comet 9P/Tempel 1 while the mother Flyby spacecraft, Hubble, and various ground observatories watched the event.

Traditional ground-based opnav (Owen *et al.* 2006) was used in the two months leading up to encounter, with pictures gradually increasing in frequency. During the last week we obtained an average of 20 pictures each hour, a group of three every nine minutes. The data volume and quick turnaround time led us to borrow the pattern-matching algorithms from our ground-based astrometry program. Still, Nick Mastrodemos, Mike Wang, Brian Rush, and Bill Owen worked around the clock in 12-h shifts during that last week, making hourly deliveries to the OD team, and we borrowed Steve Gillam from Cassini to keep watch over things the last day. The dense data set allowed us to detect the comet's rotation both astrometrically and photometrically, and the photometry was able to separate the light from the nucleus from that of the coma. This finding in turn let us determine, in cooperation with the science team, the surface brightness of the nucleus, which in turn determined the exposure times during the flyby.

Autonav (Mastrodemos *et al.* 2006) was initialized on both spacecraft two hours before impact. On the Impactor, it commanded pictures, performed OD, and computed and executed three targeting maneuvers in the last 90 min. Centerfinding used a "blobber" technique which was designed to determine the geometric center of the largest contiguous block of pixels brighter than a threshold. As the Impactor

was required to hit a lit portion of the nucleus, whose size and shape were not well known, this relatively unsophisticated technique was exactly what we wanted: it was difficult to fool, relatively impervious to noise, and it would never (well, almost never) fail to find a desirable spot on the surface.

The same Autonav software ran in parallel on the Flyby spacecraft. It operated in a different mode, concentrating on determining the time of closest approach to the comet, so that the Flyby's cameras could take pictures of the nucleus before, during, and after the impact.

The spectacular images attest to Autonav's success.

2.7 Flight Hardware

Usually optical navigation uses whatever science instruments are already on the spacecraft, but in 2005 the Mars Reconnaissance Orbiter carried a "navcam" which was designed and built for optical navigation. Steve Synnott was the principal investigator (PI) for the experiment. This small camera has a 1024×1024 CCD detector with 24 μrad (about $5''$) per pixel and a field of view of about $1.4°$. Two-second exposures show usable stars as faint as magnitude 11.0. The camera took 480 pictures of Phobos and Deimos, with Mars carefully placed outside the field of view, between February 10 and March 7, 2006. We processed the pictures on the ground in the usual fashion, and the navigation team used the optical data to refine the B-plane estimate on approach to Mars.

2.8 Development of Enabling Technologies

The past 40 years have seen immense advances in all sorts of technologies, many of which have had direct bearing on both the practice and the performance of optical navigation. Increases in computing capability (memory, speed, and data storage) have of course led to corresponding increases in opnav capability, but other technologies have advanced as well.

2.8.1 Computers

Mariner and Viking were navigated on IBM 7094 and Univac 1108 mainframe computers whose performance was wonderful then but would be laughable by modern standards. The desire to see and manipulate images on a display, not to mention the desire to minimize computing costs, led the navigation section to obtain two minicomputers, to be dedicated to optical navigation, in the 1970s. These 16-bit Modcomp IV machines came with frame buffers and TV monitors,

and our first decent image display software was written for them, partly in Fortran but mostly in assembly language. The software consisted of a multitude of small utility programs, one for each desired function. Even so, processing time was on the order of 5–10 min per picture, an improvement over Mariner and Viking. (The rest of the navigation team continued to use mainframes.)

The Modcomps lasted until 1986. We purchased some VAX minicomputers (again, with a few frame buffers and monitors) in the early 1980s, and the opnav software was running on them in time for the Uranus encounter in 1986. We had to tell the Voyager project that the Modcomps were prime and the VAXes were backup, even though every file delivery originated on a VAX. Sure enough, as we were preparing the final, critical pre-encounter delivery on a Modcomp, with a VAX running in parallel, the Modcomp crashed and never came back to life. Thank goodness for backups.

In the early 1990s we bought our first Unix workstations. Most of the Fortran code ported easily, but we had to rewrite the image display part of ONIPS, this time in C with calls to X Windows functions. That program, now running on Linux boxes, survives to this day.

2.8.2 Detectors

Cameras for Mariner, Viking, and even Voyager used vidicon detectors. These were essentially slow-scan television cameras. The anode of the vidicon tube was placed at the camera's focal plane. During an exposure, incident photons would liberate electrons, creating a localized pocket of charge. At the end of the exposure, a raster-scanning electron beam would neutralize the charge, and the fraction of the beam *not* absorbed carried the signal. The scanning beam was subject to "beam bending" distortions caused by charge buildup in the unscanned region of the picture. The edges and corners of the anode would also build up unneutralized charge, introducing additional distortion. Consequently these cameras were made with reseaux, neutral dots, or lines on the anode itself which would show up black in the images. As the positions of the reseaux were well known, having been measured during assembly, one could measure their positions in the received pictures and rectify the images.

Galileo was the first JPL deep-space mission to carry a camera with a CCD. The pixel boundaries are determined by the various epitaxial layers applied to the front surface of the CCD chip. There is therefore no distortion of the sort that plagued vidicons. However, if too much light falls on one pixel, the photoelectrons can spill over or "bleed" into adjacent pixels, rendering that image useless for photometry but only the corrupted pixels useless for astrometry.

CCDs, with their greater geometric fidelity, produce much better astrometry than the vidicons. Well-exposed star images, with an acceptable signal-to-noise

ratio, can often have rms residuals less than 0.05 pixel. The vidicons rarely gave us better than 0.25 pixel. The same results hold for unresolved images of bodies. Not only have CCDs dramatically improved opnav performance, they have also done wonders for groundbased astrometry.

2.9 Star Catalogs

In the 1960s, the best whole-sky star catalog was the one compiled by the Smithsonian Astrophysical Observatory (SAO 1966). Most of the material that went into the Smithsonian Astrophysical Observatory (SAO) catalog came from the Yale Zone Catalogs of the 1930s. The positions were already 30 years out of date, and errors in proper motion (both random errors and systematic) propagated for 30 years undoubtedly inflated the opnav residual statistics. Nevertheless, the SAO catalog served as the reference for the Mariner and Viking missions.

Beginning with Voyager, JPL contracted with Lick Observatory for the creation of *ad hoc* star catalogs. Arnold Klemola and his colleagues would point their 20-inch dual astrograph (one lens corrected for yellow light, one for blue) at coordinates provided by JPL and expose 17 × 17 inch glass photographic plates. JPLers – at various times Morabito, Taraji, Ocampo, and Owen – would travel to Santa Cruz and use Lick's survey machine to select the stars to be measured. This was a tedious process involving a screen (with crosshairs) showing a small part of the plate, a joystick and a "Record" button. Lick's automated measuring engine, geometrically a close cousin to the survey machine but with a PDP-8 in charge, would obtain precise (x, y) coordinates for each image. Klemola would then reduce the data and send JPL a catalog on magnetic tape. We used Lick's services for all the Voyager encounters, for Galileo's flybys of asteroids Gaspra and Ida, and even for the star fields behind those two asteroids as seen from the ground. This work was necessary; it turned out that some of the measured positions differed by over $2''$ (one Voyager pixel) from the positions in the old SAO catalog. Worse, the SAO had systematic errors which, if left undetected, would have produced navigation errors (in particular, errors in the B plane) that could have compromised the encounters.

The release of European Space Agency (ESA's) HIPPARCOS and Tycho catalogs (ESA 1997) changed the world of stellar astrometry overnight. Suddenly we knew the positions and proper motions of 120,000 stars to about 1 milliarcsecond (mas) and 1 mas/year, and data for another million stars to lesser accuracy. Norbert Zacharias and his colleagues at the U.S. Naval Observatory then remeasured and rereduced the Astrographic Catalogue material from the turn of the twentieth century. They also imaged the whole sky with CCD cameras, and combined the two sources to produce five releases of the USNO CCD Astrograph Catalog

("UCAC"). Version 2 of that catalog (Zacharias 2004), covering 86% of the sky (not the north polar regions), was used by Deep Impact and Cassini. Versions 4 and 5 are now in common use, although these are being superseded by ESA's Gaia catalog (Gaia Collaboration 2018).

2.10 Stereophotoclinometry

The next major advance in optical navigation techniques at JPL was the Dawn mission to orbit asteroids (4) Vesta and (1) Ceres. For both objects, opnav pictures were taken on approach and in high- and low-altitude mapping orbits. Rather than measuring landmarks by hand, as was done for NEAR, we used a mapping technique known as stereophotoclinometry (Gaskell, Barnouin-Jha and Scheeres 2007, Gaskell 2011) to model the terrain surrounding the landmarks. These maps, created from a set of images taken under varying lighting and viewing angles, can then be resampled and reprojected to match an incoming picture; the results are correlated with the new picture to determine the landmark coordinates. The technique can also be used to create global maps, and indeed the Dawn optical navigators produced global maps at increasingly finer resolution as the spacecraft drew closer to the body (Mastrodemos *et al.* 2012).

Stereophotoclinometry has since been used at JPL to support ESA's Rosetta mission to comet 67P/Churyumov–Gerisamenko and Goddard's OSIRIS–REx mission to asteroid (101955) Bennu: in the former case as a check on ESA's own navigation capabilities, and in the latter to provide an accurate shape and surface model to the project.

2.11 Future Missions

As of the publication date of this book, the only approved JPL missions that are likely to require optical navigation are the Psyche mission to explore the metallic asteroid of the same name and the Double Asteroid Redirection Test to (65803) Didymos and its satellite. We expect to use "classical opnav" using the centers of the bodies on approach, and then for Psyche to switch to landmark tracking with stereophotoclinometry. There is also a possibility that the Europa Clipper will consider opnav.

Missions to return to Uranus or Neptune have been proposed. Should these fly, optical navigation would certainly be needed to improve the orbits of the satellites, not to mention the planet-relative trajectory of the spacecraft. Any mission to an asteroid, including sample returns – and many such missions have been proposed – is likely to use opnav on approach and landmark tracking in orbit.

2.12 Optical Navigation Outside JPL

JPL's monopoly on optical navigation was broken when the Applied Physics Laboratory selected KinetX, Inc., to supply navigation services for the New Horizons mission to Pluto. Former JPLer Philip Dumont and former JPL intern Coralie Jackman put together KXIMP (pronounced "chimp"), a capable, modern system for image prediction and processing. This system performed admirably at Pluto, producing results that were in good agreement with those obtained at JPL.

KXIMP was augmented after Pluto to handle the extremely faint images of Kuiper Belt object (486958) Arrokoth,[2] the target of New Horizons' extended mission. Once again the KinetX results were in statistical agreement with the JPL results.

KinetX's involvement with optical navigation continues with their support of OSIRIS–REx, NASA's first asteroid sample return mission (Williams *et al.* 2018).

2.13 Summary

Any spacecraft which encounters an asteroid or a comet, whether it be the primary target or merely a flyby en route to something else, will probably need optical navigation, on the ground if not onboard, simply because opnav can determine the target-relative position so well, at least in two dimensions. The third dimension, namely the downtrack direction (which provides the time of closest approach), is observable in opnav imaging only after the direction to the target has moved significantly.

Missions which will require a spacecraft to be very close to a surface – landers and "touch-and-go" sample return missions – will likely require optical techniques. Whether the goal is to avoid obstacles at the landing site or to arrive near something already on the ground (a lunar base, for instance), some form of terrain-relative navigation will be necessary, and optical data will undoubtedly be a major part of it. Autonomy will probably play a huge role as well.

Flagship missions to the outer planets and their satellites will also benefit from opnav. The same need that Voyager, Galileo, and Cassini had will still apply to these missions.

Whenever there is a mission for which radio-only OD isn't sufficient, Opnav will be there. Whenever there is a mission for which the OD requirements can't be met from the ground, Autonav will be there.

---

2 The provisional designation of this object was 2014 MU_{69}, while during the encounter the New Horizons team members referred to it colloquially as "Ultima Thule."

3

Cameras

In everyday life, a *camera* is some piece of hardware whose function is to take *pictures*. Cameras come in all shapes and sizes, from the miniature ones in cell phones to the expensive equipment used by professional photographers. Cameras on spacecraft likewise span a range of sizes, from the telescope-sized narrow-angle cameras of Voyager, Galileo, and Cassini to the small detectors on cubesats. Regardless of their shape or size, all cameras share these properties:

- They have optics capable of producing a *real image* on a *focal plane*. The optics may be (in theory) as simple as a pinhole or as complicated as a many-element zoom lens. The optical system may comprise mirrors, lenses, or both.
- They have a *detector*, ideally in the focal plane of the optics, to capture the light impinging on it. The output of the detector is a two-dimensional record of the amount of light reaching the detector during an *exposure*. This two-dimensional record is a *picture*.

There is no reason why optical navigation should use terminology that departs from the vocabulary of everyday life. Cameras take pictures of a *scene*, and the resulting picture contains *images* of the *objects* in the scene.

Despite the foregoing highlighted words, one will hear of a "frame" (taken by a "framing camera") in lieu of "picture." This term was coined in order to distinguish a typical camera from a "pushbroom" camera which is continuously imaging and reading out while the scene changes. These pushbroom cameras are especially attractive when a spacecraft is in orbit, for the motion of the spacecraft relative to the surface of (say) Mars or Mercury means that the camera will sweep across a long swath of the surface, hence the name. Pushbroom cameras are seldom appropriate instruments for optical navigation, for each row of a pushbroom picture is taken at a different time. The camera pointing and spacecraft-to-surface vector are therefore different for each row. To treat a collection of rows as if it were one picture ignores these subtle effects, creates a "picture" subject to random distortions, and can easily produce erroneous results.

Spacecraft Optical Navigation, First Edition. William M. Owen Jr.

This chapter presents primarily *geometric optics*, that is, the locations of images in the focal plane of a camera.

3.1 The Gnomonic Projection

The *gnomon* of a sundial is the slanted edge—and, by extension, the supporting structure as well—that rises above the dial itself. The shadow cast by the gnomon on the dial indicates the hour angle of the apparent sun and therefore indicates the local apparent solar time. The edge of the shadow is formed by drawing a line from the sun to each point on the gnomon and then projecting that line onto the plane of the dial. The salient feature of this *gnomonic projection* is that a ray emanating from anywhere passes through one point and continues on to intersect, or impinge upon, a plane. In the case of a sundial, the rays emanate from the sun, pass the edge of the gnomon, and impinge upon the dial, thereby casting a shadow of the gnomon onto the dial. As the sun moves across the sky during the course of a day, the shadow moves too, and the dial contains markings that show the location of this shadow, hour by hour.

An ideal camera uses the same projection. To first order, one can model the lens assembly as a pinhole. A ray emanating from any point in the scene being imaged will pass through the pinhole and proceed to impinge on the plane containing a detector.

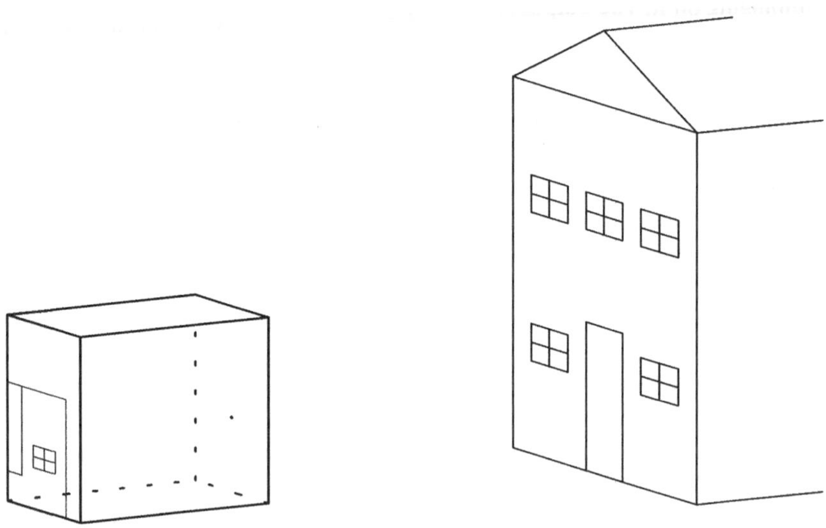

Figure 3.1 A camera obscura creating an image of the outside scene against the back wall.

The very word "camera" (Latin and Italian for "room" or "chamber") hearkens to the "camera obscura," a darkened room in which a pinhole aperture in one wall would produce on the opposite wall an upside-down image of the scene outside the room. Figure 3.1 is a crude drawing of a camera obscura, showing how part of the scene outside the building would appear to someone inside the room.

The camera coordinate system is defined by the three axes N, M, and L, with L being the optical axis of the camera. The vector $\mathbf{A}$ emanates from the origin (i.e. the pinhole) and points toward a star—more accurately, in the direction from which light from a star appears to emanate. Starlight goes through the origin and continues in a straight line until it impinges on the detector, which is placed a distance f behind the pinhole.

3.2 Deviations from the Gnomonic Projection

For most cameras the axis of symmetry of the optics (the "optical axis") is intended to pierce the center of the detector, and the detector itself is intended to be perpendicular to this axis. Neither of these intentions is necessarily achieved perfectly in practice. Furthermore, the gnomonic projection is an ideal that can be achieved only by an infinitesimally small pinhole. Real optics, whether reflective (using mirrors) or refractive (using lenses), by their very nature do not and cannot produce a perfect image. Departures of the actual image from the ideal one are historically termed *aberrations*.

3.2.1 Optical Aberrations

Textbooks such as the classic *Fundamentals of Optics* (Jenkins and White 1976) teach that lenses and mirrors are subject to five "third-order" optical aberrations, so called because $\sin\theta \neq \theta$ and the next term in the Taylor expansion is $\theta^3/3!$ These five aberrations are:

1. *Spherical aberration.* In which incoming light rays near the edges of the aperture are not focused at the same distance from the optics as those near the center of the aperture. This effect thus causes a point source to produce a blurry image. The location of the image in the focal plane is not affected.
2. *Coma.* In which off-axis rays which enter the aperture progressively farther from its center intersect the focal plane also progressively farther from the *principal point*. The result is that a bright star far from the center of the field of view will develop a shape somewhat like a badminton birdie seen from the side, with a bright spot where the image *should* be and a "tail" pointing away from the center. Early observers likened the appearance of such an image to the head

and tail of a comet, hence the name. ("Coma" is the Latin word for "hair," as in the constellation Coma Berenices.) A simple paraboloidal primary mirror, as in a Newtonian telescope, will suffer from coma.

3. *Curvature of field.* In which the best focus lies not in a plane but on a section of a sphere. If the detector is planar, images near the center may be in focus but those in the corner will not be. Classical Schmidt telescopes are known for this aberration. Indeed, the photographic glass plates used by the 48-inch Schmidt telescope at Palomar were thinner than usual, so that they could be warped to fit into the curved plate holder.
4. *Astigmatism.* Which changes out-of-focus star images from the ideal circles into ellipses. Images in focus, or very nearly so, are not affected. This aberration is most pronounced when curvature of field is also present.
5. *Distortion.* Which moves star images radially by an amount proportional to the cube of the angle of incidence. The motion can be either inward (*barrel distortion*) or outward (*pincushion distortion*), so named according to the appearance of a square grid of images. Pure distortion does not change the appearance of a point source image, but an extended body will indeed appear distorted.

3.2.2 Keystone or "Tip/Tilt" Distortion

The ideal detector lies in a plane perpendicular to the optical axis. If the detector is mounted incorrectly, at an angle to the correct plane, one side of the detector will lie closer to the aperture than the other. When this occurs, images on the nearer side will appear closer together, while those on the farther side will be spread apart. The image of a square is therefore transformed into a trapezoid. This phenomenon is a natural consequence of the gnomonic projection; it is not intrinsic to the optics.

3.3 The Creation of a Digital Picture

Beginning with the Galileo mission to Jupiter, launched in 1989, cameras on spacecraft have used solid-state detectors, originally charge-coupled devices (CCD) and more recently *complementary metal-oxide semiconductors* (CMOS), also known as active pixel sensor (APS). Earlier missions, including the Mariners, the Viking orbiters, and the Voyagers, used television cameras. The mechanisms and electronics are quite different, but the principle is the same: an incoming photon impinges upon a detector, possibly freeing an electron, at the end of the exposure these *photoelectrons* are counted, pixel by pixel, and these counts are converted into an array of digital *data numbers* or DNs, one value for each pixel. The DN array is the digital picture.

3.3.1 Vidicon Detectors

Although Vidicon detectors ("television cameras") are no longer used, the pictures they took remain useful, and therefore it is appropriate to describe these cameras briefly here.

A Vidicon is a vacuum tube. One end is glass, with a light-sensitive metal anode behind the glass. Light hitting the anode liberates an electron from the anode. The electron drains to ground, leaving a positive charge behind. At the end of the exposure, a beam of electrons from the cathode (hence the name "cathode-ray tube" for components of old televisions and monitors) sweeps across the anode in a raster pattern, which is produced by magnetic deflection coils surrounding the tube itself. The electron beam neutralizes the built-up charge on the anode. The leftover electrons produce an output voltage which in turn is inverted, measured, and digitized to produce the output DN. The inversion is necessary because bright areas on the image absorb more electrons during neutralization.

Figure 3.2 is a typical Voyager 2 opnav image of Neptune. One can see these idiosyncrasies easily:

Figure 3.2 Instrumental effects in a Vidicon picture. This is a long exposure of Neptune from Voyager 2's narrow-angle camera. Triton is the bright object at lower right. Six stars are also visible.

1. The charge in unscanned regions deflects the electron beam, so that the scans become warped. The region immediately above Neptune and Triton is artificially dark, because the electron beam has been deflected into areas of the anode that had already been scanned.
2. For the same reason, the corners appear "pinched." Charge has built up in the unscanned area of the anode surrounding the picture, and this extra charge keeps the scan from going all the way into the nominal corners of the picture.
3. Electrons have diffused into the overexposed images of Neptune and Triton, leaving holes behind. These holes are no different from those produced by photons, and they produce what seems to be a halo of light particularly around the bottom half of the images.
4. Many of the bright spots in the pictures are "blemishes" or "hot spots" that appear in every picture. These are produced by cosmic rays that have damaged the anode.
5. The dark dots and the dark bars in the corners are "reseaux" (from the French for "network"), caused by light-insensitive material placed on the anode during production in order to enable rectification of the image. One would measure the actual location of the dots surrounding an image, compare their observed location to their known location on the anode, and use the results to correct the measured location of the image.

3.3.2 CCD Detectors

The charge-coupled device, universally known as a CCD, has been the workhorse detector for both space-based and ground-based astronomical cameras since its invention around 1970. Galileo and Hubble's first Wide Field/Planetary Camera carried the first CCD cameras into space. These had 800×800 pixels, because those were the dimensions of Voyager's Vidicon detectors and Galileo could inherit much of Voyager's image processing software.

CCDs have grown larger since then, both in the number of pixels and in the physical size of the detector. Spacecraft cameras with 1024×1024 pixels are common, and commercial cameras for astronomical use are readily available with 4096×4096 pixels (Janesick *et al.* 1990).

The electronics within a CCD are described at great length in books for both amateur and professional astronomers, for instance Howell (2006) or Woodhouse (2017). While optical navigation analysts may not need to be intimately familiar with all of the workings of a CCD, it is important to understand not only how a picture is created and read out but also some of the idiosyncrasies that may cause the received picture not to be a faithful representation of the incoming light.

A CCD chip is a solid-state detector, comprising a silicon substrate with *gates* of positively or negatively doped silicon laid down on the front surface. When a photon impinges on the CCD, its energy will usually (with quantum efficiency often exceeding 90%) promote one of the electrons in a silicon atom into the *conduction band*, where they are free to move. A positively charged *hole* is left behind, and the holes too are free to move, in a manner of speaking, because other electrons can fill the void left behind.

The gates, however, can carry a voltage, and negatively charged gates form a potential barrier that restricts the motion of electrons. In this way two pairs of parallel gates, one pair vertical and the other horizontal, define the boundaries of each pixel. This scheme ensures that—up to a point—electrons that are generated within a pixel will stay within that pixel. This scheme holds until there are so many electrons within a pixel that the negative voltage of the gate structure is no longer sufficient to repel them, and then the excess charge will overflow, usually vertically in a picture, to neighboring pixels.

Once an exposure is complete, the accumulated charge in each pixel must be ascertained. In the first and simplest method, the voltage of the horizontal gates is changed methodically to shepherd the charge in each pixel into the next lower row, and the charge in the first row is transferred into a light-insensitive *readout register*. Gates within the readout register then "clock" the charge similarly, into the next lower column. The charge in the first column is deposited into a region (usually called a *port*) whose voltage changes in response to the added charge. The voltage is then amplified, sensed, digitized, and reported as a DN value. In this way a picture is read out, pixel by pixel, row by row.

Larger CCD chips can have more than one port, each responsible for a subset of the pixels. The ports are read concurrently, and the full picture is then assembled by software or firmware.

3.3.3 Active Pixel Sensors

An "APS" is similar to of a CCD in that it is a silicon-based detector with a gate structure, but the readout technique is quite different. The pixels are individually addressable, and the additional electronics required to accomplish this feat can render areas of each pixel less sensitive to light. This problem is ameliorated by microlenses, one above each pixel, that focus all of the light landing on that pixel into its light-sensitive region. Because most cell phone cameras use this technology, the use of APS detectors in spacecraft cameras has become increasingly common.

3.4 Picture Flattening

The DN in each pixel is ideally the sum of (i) a constant *bias*, applied to ensure that the output voltage remains negative, and (ii) a value directly proportional to the number of electrons. The constant of proportionality is the *gain*, traditionally expressed as electrons per DN. We thus have, for the ideal detector,

$$DN = n_e/g + b, \tag{3.1}$$

where n_e is the number of electrons, g is the gain, and b is the bias.

Real CCDs or CMOS chips differ from the ideal in several important respects. First, the conversion of voltage to DN is not exact: there is a random *read noise* (measured in either electrons or DN) present in each pixel. Second, holes and electrons can recombine during an exposure, causing the reported DN to be less than it should have been. Third, as mentioned above, when too many photons land on the same pixel, electrons may bleed into adjacent pixels in the same column. Finally, charged particles, usually protons from the solar wind, may become lodged inside the detector. These provide a continual source of electrons, and the *hot pixels* that contain them appear as permanent bright spots in each picture.

The read noise, gain, and bias are determined in the laboratory from a set of calibration images. First, the chip is repeatedly read out with the shutter closed (or, if the camera lacks a shutter, one places the camera in a darkroom and uses the shortest possible exposure). This procedure produces a set of N *bias pictures* to which the only contributions should be the bias and the inevitable read noise. These are averaged, thus decreasing the read noise by a factor $\sqrt{N}$, to produce a *master bias* picture. For opnav use, it is often sufficient simply to calculate the mean DN value, after discarding any pixels with anomalous values, and to consider that mean as the bias for the entire picture. The pixel-to-pixel variation in one of these bias pictures is the read noise. This can be calculated for each picture separately and the results averaged, yielding a representative value.

Next, the camera is aimed at a uniformly illuminated featureless target that will fill the field of view, and more pairs of pictures are taken and averaged, each pair with progressively longer exposures. Ideally the resulting pictures will be uniformly bright, with a pixel-to-pixel variation that is a combination of the read noise, the Poisson statistics that govern the number of photoelectrons generated in each pixel, and a possible *fixed pattern noise* arising from nonuniformity in the pixels' response to light.

Taking the difference between the two exposures in each pair will effectively remove the fixed pattern noise, and the scatter that remains is caused only by the read noise and the Poisson statistics. The variance of the latter is equal to the number of photons (or the standard deviation is the square root of the number of photons), and by comparing this standard deviation to the mean value *both measured in DN* one can easily obtain the *gain*, the conversion factor usually expressed as electrons per DN. This procedure is usually performed by the camera manufacturer, and optical navigators can often use the quoted results without further ado.

APS chips that were originally designed for cell phone use have interesting features that produce equally interesting features in the pictures they take. A color camera has a filter over each pixel, and the pixels are typically arranged in 2×2 blocks, with the four filters in each block in a "Bayer pattern," two green pixels (because the human eye is most sensitive to green light), one red pixel and one blue pixel. The four pixels in each block are sent to four different readout channels, with individually adjustable electronics. In this way a proper color balance can be achieved. All this is vitally important to the performance of a cell phone camera. Put one of these inside a spacecraft camera, however, and the optical navigation analyst will see artifacts peculiar to this kind of chip. The average bias level is likely to be different in each pixel within each block, and indeed we have also seen differences between even and odd *pairs* of columns. That is, for bias removal purposes the fundamental block repeats over 4 columns $\times$ 2 rows instead of the 2×2 pattern one would expect. The gain is likely to differ as well.

3.5 Readout Smear

Recent years have seen a trend toward smaller spacecraft and consequently smaller cameras. One consequence of this trend is the elimination of a mechanical shutter, which admits light into the camera only when it is open, in favor of a so-called "electronic shutter." At the beginning of an exposure, the chip (either CCD or CMOS) is rapidly flushed: the accumulated charge is clocked out, but it is allowed to drain to ground rather than being read. At the end of the exposure, the charge in each pixel is transferred quickly into a storage area, not exposed to light, and then read out pixel by pixel as usual. During these two operations—the initial flush and the final transfer—light from the scene being imaged continues to fall onto the chip. As the transfers are performed in the column direction, light from a bright object in the picture will contaminate all the pixels in the columns in which that object appears. This phenomenon is "readout smear." It doesn't actually *smear* the image itself, in the same way that a moving camera would. It simply produces additional DN in any column containing a sufficiently bright image.

In "one-sided" smear, the initial flush is instantaneous but the final transfer is not. Let the exposure duration be T and the time required to shift the image by *one row* during readout be t_r. The readout direction is toward lower row numbers, because the first row to be read out is by definition row 1, the top row. This row will be uncontaminated. Row 2 is shifted into row 1 while row 1 is being read out, and during this time it will gain, in addition to the charge it received during the exposure, a fraction t_r/T of the contents of row 1. So throughout the picture, each row receives during readout a contribution from every row with lower row numbers. The effect is that any bright image will produce a "tail" in the rows with higher row numbers.

A method for removing one-sided smear was published by Wim Ruyten (1999). As stated above, the top row (row 0 for a C programmer, row 1 in Fortran) is unaffected by readout smear. The next row is affected only by the top row, so if one subtracts (t_r/T) times the top row from it, this subtraction will remove the readout smear from the next row. Now that the first two rows have been reconstituted, one subtracts $(t_r - T)$ times *both* the first and second row from the third row. The process is repeated, row by row, throughout the picture. Let r and c be the row and column numbers (beginning with 0), **N** be the observed DN array, and **B** be the true brightness array. We can recover **B** with Ruyten's algorithm. In C, where the top row is number 0:

```
ρ = t_r/T
for (c = 0 to n_c − 1) { /* outer loop over columns */
  B_{0,c} = N_{0,c}     /* top row is unchanged */
  s = ρ B_{0,c}     /* smear contribution from top row */
  for (r = 1 to n_r − 1) { /* inner loop over rows in this column */
      B_{r,c} = N_{r,c} − s /* correct this pixel */
      s = s + ρ B_{r,c} /* add its contribution to smear */
  }
}
```

This algorithm may be performed "in place," that is, with **B** and **N** being the same array.

"Two-sided" smear (Figure 3.3) arises when both the initial flush and final transfer occur in finite time. The removal of two-sided smear has historically been a difficult problem to solve. As the signal in each pixel now includes contributions from all other pixels in its column, one can express the problem in matrix form. It is clear that

$$\mathbf{N} = \begin{pmatrix} 1 & r & r & \dots & r \\ f & 1 & r & \dots & r \\ f & f & 1 & \dots & r \\ \vdots & \vdots & \vdots & \ddots & \vdots \\ f & f & f & \dots & 1 \end{pmatrix} \mathbf{B}, \tag{3.2}$$

where r and f are the ratios of the readout and flush times, t_r and t_f, respectively, to the exposure time t_e.

A brute-force approach would simply invert this matrix to solve for **B**. This matrix has as many rows and columns as there are rows in the detector, and it is not fast to invert a 1024×1024 matrix. Cheng *et al.* (2008) presented an iterative algorithm used on pictures from New Horizons' long-range reconnaissance imager (LORRi) instrument, for which $t_f \neq t_r$.

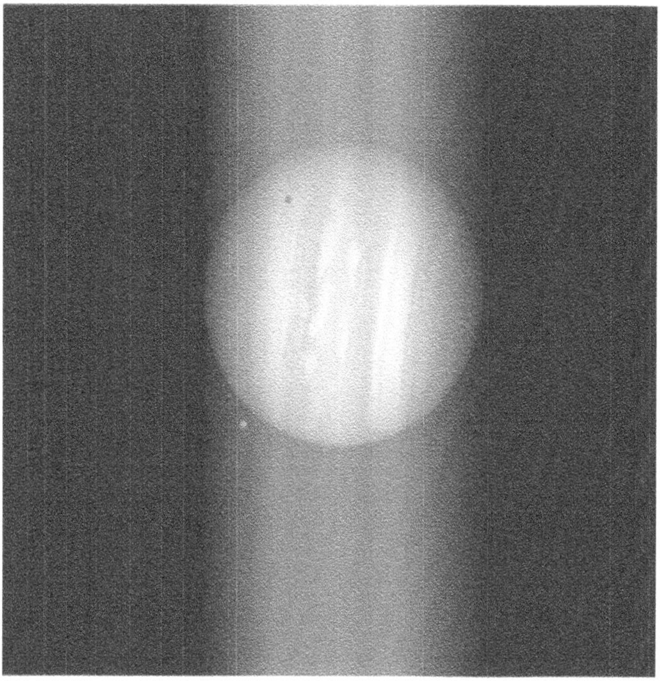

Figure 3.3 New Horizons picture of Jupiter, showing the effects of two-sided readout smear (Owen, Weaver and Cheng 2019).

A fast algorithm (Owen, Weaver and Cheng 2019) does exist for the special case where $t_f = t_r$. Define $\epsilon \equiv t_f/t_e$, and let R be the number of rows in the detector. The total observed DN in any pixel is given by

$$N_{ij} = B_{ij} + \epsilon \sum_{k \neq j} B_{ik} \tag{3.3}$$

$$= (1 - \epsilon)B_{ij} + \epsilon \sum_{k} B_{ik} \tag{3.4}$$

$$\equiv (1 - \epsilon)B_{ij} + \epsilon \beta_i, \tag{3.5}$$

where β for column i is the sum of *all* of the (unknown) B_{ij}. We can sum Eq. (3.5) over j to get an expression for ν_i, the sum of the observed brightness in column i:

$$\nu_i \equiv \sum_{j=1}^{R} N_{ij} \tag{3.6}$$

$$= (1 - \epsilon)\beta_i + R\epsilon\beta_i \tag{3.7}$$

$$= (1 + (R - 1)\epsilon)\beta_i. \tag{3.8}$$

Figure 3.4 The same picture as in Figure 3.3, after desmearing.

That is, the sum of the observed DN in a column is a factor of $(1 + (R - 1)\epsilon)$ greater than the sum of the unsmeared DN, and this factor is known. More important, however, is the *difference* Δ_i between the observed and unsmeared sums (Figures 3.4 and 3.5):

$$\Delta_i \equiv \nu_i - \beta_i \tag{3.9}$$

$$= \nu_i \frac{(R-1)\epsilon}{1+(R-1)\epsilon}. \tag{3.10}$$

We also know, from summing the second term on the right-hand side of Eq. (3.5) over j, that

$$\Delta_i = (R-1)\epsilon\beta_i. \tag{3.11}$$

The factor $(R - 1)$ arises because there is no contribution to the readout smear in any pixel from the light landing on that pixel. If we divide Eq. (3.11) by $(R - 1)$ we must therefore get the gross contribution δ_i to readout smear from *all* pixels in the column:

$$\delta_i = \frac{\nu_i \epsilon}{1+(R-1)\epsilon} = \epsilon\beta_i. \tag{3.12}$$

Figure 3.5 The same picture as in Figure 3.3, after desmearing, at high contrast.

If we subtract this δ_i from the observed v_{ij}, the difference is almost the desired B_{ij}—not exactly, because ϵB_{ij} (smear contribution from the pixel in question) has been subtracted when it ought not to have been. Therefore we have

$$v_{ij} - \delta_i = v_{ij} - \epsilon\beta_i = B_{ij} - \epsilon B_{ij}, \tag{3.13}$$

from which the final result is quickly derived:

$$B_{ij} = \frac{v_{ij} - \delta_i}{1 - \epsilon}. \tag{3.14}$$

If $t_f \neq t_r$, one can first perform the two-sided smear removal as described above and follow that with Ruyten's algorithm using $(t_r - t_f)$ in place of t_f. This technique works even if $t_f > t_r$, for in that case Luyten's algorithm will restore the readout smear that had been overcorrected by the two-sided algorithm. The next two figures show the same picture after smear removal, first with the same contrast as the original and then to a much higher contrast. The "salt and pepper" appearance of the background above and below Jupiter arises from the increased shot noise there.

4

Modeling Optical Navigation Observables

4.1 Introduction

This chapter describes the process of predicting the image coordinates of a *target* as seen by a camera. The target may be a star, the center of some natural or artificial body, some identifiable feature or *landmark* on a body, or a point on the limb or terminator of a body. The camera is of course some device that projects the scene in front of it onto a detector of some kind, thereby forming an *image* of the scene on the detector. Light (perhaps reflected sunlight) emanates from the target, enters the aperture of a camera, and impinges on its detector, thereby forming an image. The location of the image depends on the direction of the incoming light, as expressed in a camera-fixed coordinate system, and on the optical properties of the camera.

Because light travels at a finite speed, it is not enough to know the position of the target relative to the camera at the time of the exposure. The camera sees the target, not as it is at the exposure time, but as it appeared when the light left it. One therefore must compute the target's position at a *retarded time*.

The direction of the incoming light is also changed because of the velocity of the camera itself. Although this phenomenon is commonly termed *stellar aberration* because its effects were first noticed for stars, it affects all incoming light regardless of the source.

Given the direction of the incoming light, the particulars of the camera optics and detector geometry determine where the light will impinge on the detector. The layout of the pixels then allows one to translate this location into picture coordinates.

There are thus four steps in the prediction process:

1. Determining the apparent direction from the camera to the target at the exposure time,

Spacecraft Optical Navigation, First Edition. William M. Owen Jr.

2. Expressing this vector in the coordinate system of the camera,
3. Projecting the vector through the camera optics onto the detector, and
4. Converting the projected image location into picture coordinates.

Sections 4.2–4.5 treat the first step (finding the apparent direction) for distinct objects, for stars, and for limb or terminator points on objects. Section 4.6 covers the various rotations required to express this apparent direction in camera coordinates. Sections 4.7 and 4.8 recap the ideal or "gnomonic" projection and deviations from it. Section 4.9 describes the conversion into sample and line coordinates. The final section provides a summary.

4.2 The Apparent Direction to an Object

This section deals with the geometry calculations that culminate in predicting the apparent direction from the *camera* to an *object*. The camera need not be located on a spacecraft: the same technique can be used if the camera is instead a telescope at an observatory. The object can be the center of any physical body in the Solar System (including another spacecraft) or a recognizable feature or *landmark* on one of these bodies.

The time t at which the camera's location $\mathbf{R}(t)$ is computed is the midpoint of the exposure, expressed in barycentric dynamical time or TDB. In its most general form, $\mathbf{R}$ is computed as the vector sum of the camera's position with respect to its *parent* (usually but not necessarily a spacecraft), the position of that parent with respect to its own parent, etc., and the position of the final parent with respect to the Solar System Barycenter. Two examples will suffice.

Example 4.1 The camera is on a spacecraft orbiting Europa.

$$\begin{aligned} \mathbf{R} = {} & \text{camera with respect to spacecraft center of mass} \\ & + \text{spacecraft center of mass with respect to Europa center} \\ & + \text{Europa center with respect to Jupiter system barycenter} \\ & + \text{Jupiter system barycenter with respect to Solar System Barycenter.} \end{aligned}$$

The first vector might be constant *in spacecraft coordinates* such that it must be rotated into International Celestial Reference Frame (ICRF). The second could come from a Europa-centered spacecraft ephemeris file. The third and fourth could be found from a satellite ephemeris and planetary ephemeris, respectively.

Example 4.2 The camera is on the Hale telescope at Palomar Observatory.

$$\begin{aligned}\mathbf{R} = {} & \text{camera with respect to telescope center}\\ & + \text{telescope center with respect to center of Earth}\\ & + \text{center of Earth with respect to Earth} - \text{Moon barycenter}\\ & + \text{Earth} - \text{Moon barycenter with respect to Solar System Barycenter.}\end{aligned}$$

The first vector in this example may well be ignored. The second is found from the terrestrial coordinates (longitude, latitude, elevation) of the observatory and then rotated from Earth-fixed to ICRF coordinates using precession, nutation, Greenwich sidereal time, and polar motion. The third and fourth would again be found in a planetary ephemeris.

The position of the target object can similarly be related to the Solar System Barycenter.

- A *landmark* on a body has a defined position relative to the center of the body. This vector is in body-fixed coordinates and must be rotated into the ICRF coordinate frame.
- The position of the center of a body is found from an ephemeris file or from theory. In either case the vector has an origin: typically the center of a planet, the barycenter of a planetary system, the Sun, or the Solar System Barycenter.

Therefore, one begins with either the camera or something being imaged, finds its position relative to some coordinate origin, then finds that origin's position with respect to some other point, and so forth until one reaches the Solar System barycenter. The desired vector, from the Solar System barycenter to the object of interest, is the sum of all these constituent vectors. Velocity, if needed, is treated similarly.

The camera, however, does not "see" the object at its current or *geometric* position. Two physical phenomena affect what the camera detects:

1. The light that enters the camera at time t left the object at some earlier or *retarded time* $t - \tau$. The camera sees the object not as it is but as it was. This effect is called "light time," and the object's position vector at the retarded time is called its *true* position.
2. If the camera has a nonzero velocity with respect to the Solar System Barycenter, all light entering it will appear to have emanated from a direction that has been displaced toward the camera's velocity vector. This effect is known as *stellar aberration* because it was detected first in star positions, although

it affects all incoming light regardless of its source. The object's observed direction, corrected for both light time and stellar aberration, is known as its *apparent* direction.

The apparent position of the object is calculated as follows:

1. Find the camera's position $\mathbf{R}(t)$ and velocity $\dot{\mathbf{R}}(t)$ of the camera with respect to the Solar System Barycenter.
2. Find the object's geometric position $\mathbf{S}_0(t)$ with respect to the Solar System Barycenter.
3. The initial guess for the light time is

$$\tau_0 = |(S_0(t) - \mathbf{R}(t))|/c. \tag{4.1}$$

4. Find the initial guess for the object's true position $\mathbf{S}_1(t - \tau_0)$ as for $\mathbf{S}_0(t)$ but at the presumed retarded time.
5. Iterate steps 3 and 4 as necessary until $\tau_i \to \tau$. Usually two iterations suffice. Then $\mathbf{S}_n$ after the final iteration is taken to be $\mathbf{S}(t - \tau)$, the object's true position at the retarded time $(t - \tau)$.
6. The true direction $\mathbf{T}(t)$ from the camera to the object is found by a simple vector subtraction:

$$\mathbf{T}(t) = \mathbf{S}(t - \tau) - \mathbf{R}(t). \tag{4.2}$$

7. The apparent direction $\mathbf{A}(t)$ from the camera to the object is formed by correcting the true direction for stellar aberration. This correction is easily expressed in Newtonian mechanics by

$$\mathbf{A}(t) = \mathbf{T}(t) + \tau\dot{\mathbf{R}}(t). \tag{4.3}$$

The relativistic formula produces results which are not significantly different from the above, for typical spacecraft velocities, and Jet Propulsion Laboratory (JPL) does not use it.

4.3 The Apparent Direction to a Star

Star catalogs provide for some set of stars the right ascension, declination, *proper motion* (angular rates in right ascension and declination) and sometimes the *parallax* (the angle subtended by one astronomical unit at the star's distance). Catalogs typically include the *central epoch* for the right ascension and declination; this is the weighted mean of the times of observation and also the time at which the position uncertainty is minimum. (The central epochs in right ascension and in declination may differ.) Catalogs also include the uncertainty in the two components of proper motion and sometimes the uncertainty of the parallax.

The star's *magnitude* (apparent brightness), color indices (difference in magnitudes between two passbands), spectral type, and other parameters may also be catalogued. Some catalogs also include the stars' radial velocity.

Let α_0 be the star's catalogued right ascension at the central epoch t_α, μ_α be the proper motion in right ascension, and σ_{α_0} and σ_{μ_α} be the catalogued uncertainty in α_0 and μ_α, respectively. Let δ_0, t_δ, μ_δ, σ_{δ_0}, and σ_{μ_δ} be the corresponding quantities in declination. All these are nowadays given in the International Celestial Reference System (ICRS). Let π be the parallax, measured in arc seconds, and let v_r be the radial velocity, measured in km/s.

The standard practice, acceptable for most stars, is to assume that the right ascension and declination are each linear functions of time. Under this assumption, the coordinates propagate to time t by

$$\alpha(t) = \alpha_0 + (t - t_\alpha)\mu_\alpha; \tag{4.4a}$$

$$\delta(t) = \delta_0 + (t - t_\delta)\mu_\delta. \tag{4.4b}$$

The uncertainty in the spherical coordinates at time t is

$$\sigma_\alpha^2(t) = \sigma_{\alpha_0}^2 + (t - t_\alpha)^2\sigma_{\mu_\alpha}^2; \tag{4.5a}$$

$$\sigma_\delta^2(t) = \sigma_{\delta_0}^2 + (t - t_\delta)^2\sigma_{\mu_\delta}^2. \tag{4.5b}$$

If the parallax is known, the vector $\mathbf{r}(t)$ from the Solar System barycenter to the star is simply

$$\mathbf{r}(t) = \frac{1\ \text{pc}}{\pi}\begin{pmatrix} \cos\alpha(t)\cos\delta(t) \\ \sin\alpha(t)\cos\delta(t) \\ \sin\delta(t) \end{pmatrix}, \tag{4.6}$$

where 1 parsec is the distance at which 1 astronomical unit subtends an angle of 1 arc second, using the small angle approximation $\sin\theta \approx \tan\theta \approx \theta$. More rigorously (IAU 2015 General Assembly, Resolution B2),

$$\begin{aligned} 1\ \text{pc} &= (180 \cdot 3600)/3.14159\ldots\ \text{au} \\ &\approx 3.085677581491 \times 10^{13}\ \text{km}. \end{aligned} \tag{4.7}$$

If the parallax is not known, one can simply use a very large value for the magnitude of $\mathbf{r}(t)$.

Note that the light-time correction is not used for stars. The catalogued direction is already corrected for light time. The star's true position as seen from the camera, corrected for parallax, is therefore given by

$$\mathbf{T}(t) = \mathbf{r}(t) - \mathbf{R}(t), \tag{4.8}$$

where $\mathbf{R}(t)$ is, as before, the position of the camera with respect to the Solar System barycenter.

For a few stars—ones with high proper motion and ones very close to a celestial pole—the linear approximation given above is inadequate. A better model is to assume instead that the star's *space velocity* $\dot{\mathbf{r}} \equiv \mathbf{v}$ is constant; that is, that the star is in rectilinear motion relative to the solar system. Let $\mathbf{r}_0$ be the star's position vector at some time t_0. (If $t_\alpha = t_\delta$ the choice of t_0 is clear.) Let α and δ be the star's coordinates at this time. Let $\dot{r}$ be the star's radial velocity. Then the space velocity is

$$\mathbf{v} = \dot{r}\begin{pmatrix} \cos\alpha\cos\delta \\ \sin\alpha\cos\delta \\ \sin\delta \end{pmatrix} + r\mu_\alpha\cos\delta\begin{pmatrix} -\sin\alpha \\ \cos\alpha \\ 0 \end{pmatrix} + r\mu_\delta\begin{pmatrix} -\cos\alpha\sin\delta \\ -\sin\alpha\sin\delta \\ \cos\delta \end{pmatrix} \tag{4.9}$$

with the quantities expressed in consistent units: proper motions in radians/second, $\mathbf{r}$ in km, $\dot{r}$ in km/s.

Once $\mathbf{T}(t)$ is known, the apparent direction $\mathbf{A}(t)$ to the star is formed by correcting for stellar aberration. Form the unit vector $\hat{\mathbf{r}} = \mathbf{T}(t)/|\mathbf{T}(t)|$, and then

$$\mathbf{A}(t) = \hat{\mathbf{r}} + \dot{\mathbf{R}}(t)/c. \tag{4.10}$$

As before, $\dot{\mathbf{R}}(t)$ is the velocity of the camera with respect to the Solar System barycenter.

4.4 The Apparent Direction to the Limb or Terminator

The *limb* of a body consists of those points on the surface such that a line from some exterior point (for our purposes, the camera) is tangent to the surface at the limb point. The limb therefore separates the part of the surface that is facing the camera from the part that is facing away from the camera, or the front side from the back side.

Similarly, we can define the *terminator* of a body as the set of points on the surface such that the sun is bisected by the local horizon: the apparent direction $\mathbf{S}$ to the sun is normal to the surface normal at that point. Thus the terminator divides day from night on the surface (under the assumption that the sun is a point source). For an external observer, the terminator separates the visible (sunlit) surface from the invisible surface.

The shapes of the planets and most of the larger satellites can be adequately modeled as an *ellipsoid*. The surface is defined by the equation

$$\left(\frac{x}{a}\right)^2 + \left(\frac{y}{b}\right)^2 + \left(\frac{z}{c}\right)^2 = 1, \tag{4.11}$$

in a body-fixed coordinate system B whose axes are aligned with the *principal axes* of the ellipsoid and whose origin is the center of the ellipsoid. The variables a, b, and c are the lengths of the three semiaxes. The object is an *oblate spheroid* if $a = b > c$, a *prolate spheroid* if $a = b < c$, and of course a sphere if $a = b = c$. If all three axes are unequal, the object is a *triaxial ellipsoid*. We will use the general term "ellipsoid" to refer to all these cases. Many bodies in the Solar System are adequately modeled as ellipsoids.

The ellipsoid equation can be rewritten (still in body-fixed coordinates) as

$$\mathbf{p}^T\mathbf{E}\mathbf{p} = 1, \tag{4.12}$$

where $\mathbf{p} = (x, y, z)^T$ and $\mathbf{E} = \text{diag}\,(1/a^2, 1/b^2, 1/c^2)$. Note that this transformed equation can be regarded as the equation of the unit sphere, but represented in a coordinate system whose units of length are not the same along each axis: the units have been stretched or squeezed to transform the ellipsoid into a sphere.

Let $\mathbf{A}$ be the apparent vector from the camera to the center of the ellipsoid, and again let $\mathbf{p}$ be the vector from the center of the ellipsoid to some point on it. The normal $\mathbf{n}$ to the surface of the ellipsoid at $\mathbf{p}$ is given in body-fixed coordinates by

$$\mathbf{n} = \nabla\mathbf{p} = (x/a^2, y/b^2, z/c^2)^{\mathrm{T}}. \tag{4.13}$$

It is evident from the ellipsoid equation that

$$\mathbf{n} \cdot \mathbf{p} = 1. \tag{4.14}$$

Because the normal to a limb point is perpendicular to the line of sight,

$$(\mathbf{A} + \mathbf{p})^{\mathrm{T}}\ \mathbf{n} = 0 \tag{4.15}$$

or

$$\mathbf{A}^{\mathrm{T}}\ \mathbf{E}\mathbf{p} = -1. \tag{4.16}$$

For, the *terminator* let $\mathbf{S}$ be the vector from the center of the ellipsoid to the apparent sun. We can safely ignore the size of the ellipsoid compared to that of the sun and write the equation for terminator points simply as

$$\mathbf{S}^{\mathrm{T}}\mathbf{n} = 0, \tag{4.17}$$

or, using the definition of the normal,

$$\mathbf{S}^{\mathrm{T}}\,\mathbf{E}\mathbf{p} = 0. \tag{4.18}$$

The *cusps* of an illuminated ellipsoid are points where the limb and terminator intersect: the boundaries between the *lit limb* and the *dark limb*. Because the limb plane does not pass through the center of the ellipsoid, it is possible for the limb plane and terminator plane not to intersect on the ellipsoid. In this case the entire visible surface is either illuminated or in shadow.

Denote the matrix corresponding to $\mathbf{E}$ in camera coordinates by

$$\mathbf{Q} = \mathbf{T}_B^C \, \mathbf{E} \, \mathbf{T}_B^{C^{\mathrm{T}}}, \tag{4.19}$$

where $\mathbf{T}_B^C$ is the rotation matrix *from* body-fixed *to* camera coordinates. Throughout the rest of this section, we will work in camera coordinates.

It is convenient to introduce the concept of a *scan line*. In a picture, a scan line is a ray $\mathbf{s}$ emanating from the center of a body, in a direction defined by an arbitrary *scan angle*. In space, a scan line maps to a half-plane bounded by the apparent vector $\mathbf{A}$ from the camera to the center of the ellipsoid; the orientation of the half-plane is governed by the scan vector $\mathbf{s}$. A point on the ellipsoid lies in the scan plane if it has no component perpendicular to that plane (which direction is given by $\mathbf{A} \times \mathbf{s}$), or if

$$(\mathbf{A} \times \mathbf{s})^{\mathrm{T}} \, \mathbf{p} = 0. \tag{4.20}$$

Therefore, to determine the location $\mathbf{p}$ of a limb or terminator point along a given scan line, in camera coordinates, one must solve a system of one quadratic and two linear equations:

$$\mathbf{p}^{\mathrm{T}} \mathbf{Q} \, \mathbf{p} = 1; \tag{4.12}$$

$$\text{either } \mathbf{A}^{\mathrm{T}} \, \mathbf{Q} \, \mathbf{p} = -1 \text{ for limb points} \tag{4.16}$$

$$\text{or } \mathbf{S}^{\mathrm{T}} \, \mathbf{Q} \, \mathbf{p} = 0 \text{ for terminator points;} \tag{4.18}$$

$$(\mathbf{A} \times \mathbf{s})^{\mathrm{T}} \, \mathbf{p} = 0. \tag{4.20}$$

The quadratic equation is of course the equation of the ellipsoid; the two linear equations describe planes. The two planes, unless they are parallel, intersect in a line. The line, if it exists, may intersect the ellipsoid at two points, may be tangent to the ellipsoid at one point, or may not intersect the ellipsoid at all. If there are two points of intersection, one will be on the "back" side of the ellipsoid, as seen from the camera, and therefore not visible.

A reasonable method for solving these equations proceeds as follows. First, rewrite the two linear equations as

$$\boldsymbol{\alpha}^{\mathrm{T}} \mathbf{p} = A \text{ and} \tag{4.21a}$$

$$\boldsymbol{\beta}^{\mathrm{T}} \mathbf{p} = B. \tag{4.21b}$$

Then decompose $\mathbf{p}$ in terms of its components along the directions defined by $\boldsymbol{\alpha}$, $\boldsymbol{\beta}$, and their cross product:

$$\mathbf{p} = P_\alpha \boldsymbol{\alpha} + P_\beta \boldsymbol{\beta} + P_\times (\boldsymbol{\alpha} \times \boldsymbol{\beta}). \tag{4.22}$$

This is legitimate provided $\boldsymbol{\alpha} \neq \mathbf{0}$, $\boldsymbol{\beta} \neq \mathbf{0}$, and $\boldsymbol{\alpha} \times \boldsymbol{\beta} \neq \mathbf{0}$. The last of these conditions ensures that $\boldsymbol{\alpha}$ and $\boldsymbol{\beta}$ are neither parallel nor antiparallel. Should $\boldsymbol{\alpha} \times \boldsymbol{\beta} = \mathbf{0}$, then the two linear equations are either equivalent or inconsistent.

The scalar constants P_α and P_β are found by substituting Eq. (4.22) into both (4.21a) and (4.21b):

$$\mathbf{p} \cdot \boldsymbol{\alpha} = P_\alpha(\boldsymbol{\alpha} \cdot \boldsymbol{\alpha}) + P_\beta(\boldsymbol{\beta} \cdot \boldsymbol{\alpha}) + P_\times[(\boldsymbol{\alpha} \times \boldsymbol{\beta}) \cdot \boldsymbol{\alpha}]$$
$$= P_\alpha|\boldsymbol{\alpha}|^2 + P_\beta(\boldsymbol{\alpha} \cdot \boldsymbol{\beta}) = A \tag{4.23}$$
$$\mathbf{p} \cdot \boldsymbol{\beta} = P_\alpha(\boldsymbol{\alpha} \cdot \boldsymbol{\beta}) + P_\beta(\boldsymbol{\beta} \cdot \boldsymbol{\beta}) + P_\times[(\boldsymbol{\alpha} \times \boldsymbol{\beta}) \cdot \boldsymbol{\beta}]$$
$$= P_\alpha(\boldsymbol{\alpha} \cdot \boldsymbol{\beta}) + P_\beta|\boldsymbol{\beta}|^2 = B \tag{4.24}$$

yielding two linear equations for P_α and P_β, with solution

$$P_\alpha = \frac{|\boldsymbol{\beta}|^2 A - (\boldsymbol{\alpha} \cdot \boldsymbol{\beta})B}{|\boldsymbol{\alpha} \times \boldsymbol{\beta}|^2}, \tag{4.25}$$

$$P_\beta = \frac{|\boldsymbol{\alpha}|^2 B - (\boldsymbol{\alpha} \cdot \boldsymbol{\beta})A}{|\boldsymbol{\alpha} \times \boldsymbol{\beta}|^2}. \tag{4.26}$$

The value for $P_\times$ remains arbitrary, since this component does not contribute to equation (4.21a) or (4.21b).

Now define $\mathbf{U} \equiv P_\alpha\boldsymbol{\alpha} + P_\beta\boldsymbol{\beta}$, $\mathbf{V} \equiv \boldsymbol{\alpha} \times \boldsymbol{\beta}$, and $v \equiv P_\times$. Thus Eq. (4.22) becomes

$$\mathbf{P} = \mathbf{U} + v\mathbf{V}, \tag{4.27}$$

and Eq. (4.12) becomes

$$(\mathbf{U} + v\mathbf{V})^{\mathrm{T}}\ \mathbf{Q}\ (\mathbf{U} + v\mathbf{V}) = 1. \tag{4.28}$$

The only remaining unknown is v. The equation expands to

$$(\mathbf{V}^{\mathrm{T}}\ \mathbf{Q}\ \mathbf{V})v^2 + (\mathbf{U}^{\mathrm{T}}\ \mathbf{Q}\ \mathbf{V} + \mathbf{V}^{\mathrm{T}}\ \mathbf{Q}\ \mathbf{U})v + (\mathbf{U}^{\mathrm{T}}\ \mathbf{Q}\ \mathbf{U} - 1) = 0, \tag{4.29}$$

a quadratic equation in v. Since A is symmetric, so is Q, and the two terms in v are equal.

Equation (4.29) may have zero, one, or two real solutions for v, yielding through Eq. (4.27) the final solution(s) for P.

- For finding limb points, there will be two solutions, but only one of them will lie in the half-plane determined by $\mathbf{s}$ and $\mathbf{P}_0$. (The other solution lies in the other half-plane.) Choose the solution with $\mathbf{P}^{\mathrm{T}}\ \mathbf{s} > 0$, indicating that the point is sunlit.
- For finding terminator points, there will again be two solutions, of which at most one is on the part of the ellipsoid that faces toward the observer. Choose the solution (if any) with $(\mathbf{P}_0 + P)\ \mathbf{Q}\ \mathbf{P} > 0$.
- For finding cusp points, there will be no real solutions if the terminator is everywhere behind the limb: the visible portion of the ellipsoid is either completely illuminated or completely dark. If the vector from the observer to the sun is tangent to the limb, there will be a double solution at the point of tangency. Otherwise, there will be two solutions, one for each of the two cusp points.

In practice, the cusps (if they exist) are found first, and then **s** is varied through 360° of position angle (cusp 1 to cusp 2, then cusp 2 around to cusp 1) to find a discrete set of limb and terminator points.

Most asteroids and small satellites have shapes that depart significantly from a triaxial ellipsoid. The surface of such a body can be modeled in three different ways:

1. As a network of *triangular plates*. The model contains a numbered list of *vertices*, with body-fixed coordinates for each, and a list of triangles, each defined by the numbers of its three vertices. There is no restriction on the number of triangles which may share a common vertex.
2. As a network of *pseudo-rectangular plates*. Here each plate is defined by four vertices. As the four points are not necessarily coplanar, bilinear interpolation is required to find the body-fixed coordinate of a point on the surface.
3. As a sum of spherical harmonics. The radius r at longitude λ and latitude φ is modeled as

$$r(\varphi, \lambda) = \sum_{n=0}^{N} \sum_{m=0}^{\min(n,M)} C_{nm}\, P_n^m(\sin\varphi)\, \cos(m\lambda) + \sum_{n=1}^{N} \sum_{m=1}^{\min(n,M)} S_{nm}\, P_n^m(\sin\varphi)\, \sin(m\lambda), \tag{4.30}$$

where $P_n(x)$ are the Legendre functions and $P_n^m(x)$ are the associated Legendre functions. The coefficients C_{nm} and S_{nm} are the amplitudes for each of the harmonics. The faithfulness of the model improves with increasing values of the maximum degree N and order M.

The spherical harmonic model suffers from ringing near abrupt changes in topography, and as it is a single-valued function of latitude and longitude, it cannot accommodate overhangs. The plate models, by contrast, can model any shape if the plates are small enough.

Many calculations for irregular bodies must be done through brute force. To find the point where a ray intersects the surface, for instance, one steps along the ray and at each step transforms the point into body-fixed coordinates relative to the center of the body and determines whether the point lies outside or inside the body. If the point is inside the body, this point and its predecessor have bracketed the surface, and halving the step size repeatedly will converge on the surface point. To find a limb point, one performs this calculation for various directions along the scan line. These algorithms are known to be bounded, as the minimum and maximum radii are known *a priori* from the model.

4.5 The Orientation of the Camera

It is obvious from everyday life that the contents of a picture depend on where the camera is pointed—indeed, not only on the direction of the boresight but also on the rotation angle about the boresight. The rotation matrix which transforms from the fundamental coordinate system to the camera coordinate system is traditionally known as "the **C** matrix" ("C" standing for camera). The details of the construction of **C** vary from one mission to another. Some spacecraft mount their cameras on an articulating "scan platform," whereas others bolt the camera onto the spacecraft bus so that the entire spacecraft must rotate in order to point the camera properly. Even if a spacecraft can point the camera in any direction with any possible roll angle about the boresight, often other considerations dictate what that roll angle must be. For instance, the angle between the sun and the normal to the solar arrays must be minimum, or the earth must remain in the beam of a fan-beam antenna. A few possibilities, drawn from actual missions, are discussed below.

4.5.1 Three Pointing Angles Are Known

Any rotation matrix can be expressed as the product of three elementary rotations. Many missions provide three angles in the header to each picture file: the right ascension α and declination δ of the camera boresight, and some sort of twist angle ϕ about that boresight. There are two conventions for these three rotations:

1. The 3–1–3 convention, in which the declination enters as a rotation about the intermediate x-axis:

$$\mathbf{C} = \mathbf{R}_3(\phi_1)\,\mathbf{R}_1(90^\circ - \delta)\,\mathbf{R}_3(\alpha + 90^\circ). \tag{4.31}$$

   The first rotation, about the z-axis of the fundamental coordinate system, places the boresight into the $-y$-z half plane. The second rotation, about the intermediate x-axis, rotates the boresight to its correct direction. The third rotation, about the final boresight, provides the desired twist orientation of the picture. If $\phi_1 = 0$, the fundamental x-axis will project into the horizontal direction of the picture.

2. The 3–2–3 convention, in which the declination enters as a rotation about the intermediate y-axis:

$$\mathbf{C} = \mathbf{R}_3(\phi_2)\,\mathbf{R}_2(90^\circ - \delta)\,\mathbf{R}_3(\alpha). \tag{4.32}$$

   In this convention, the first rotation places the boresight into the $+x$-z half plane. The second rotation is about the intermediate y-axis, and again the third rotation accounts for the twist. In this convention, if ϕ_2 is zero, the fundamental y-axis will be horizontal in the picture.

The two twist angles ϕ_1 and ϕ_2 are related by

$$\phi_2 = \phi_1 + 90°. \tag{4.33}$$

This relationship makes intuitive sense: if $\delta = 90°$ so that the second rotation is identity, the "extra" 90° in the 3–1–3 formulation must be counteracted by a −90° twist in order to match the 3–2–3 result.

Because of this 90° difference in the twist angles, analysts must ensure that they are interpreting the angles in the picture header correctly. Examining one picture of a star field and comparing the predicted to observed star patterns will quickly reveal any erroneous assumptions.

4.5.2 Two Pointing Angles Are Known

Some spacecraft have attitude flight rules that effectively force the twist angle to take a particular value for a given boresight direction. In many cases there is a requirement that the angle between the direction to the sun and the normal to an array of solar panels be minimized. The Voyagers' cameras were mounted on a "scan platform" with two degrees of freedom, and the orientation of the spacecraft itself provided a basic coordinate system such that only two further rotations, one in "azimuth" and one in "elevation," were necessary to point the cameras. Galileo used the roll angle of the spacecraft body plus one rotation of a scan platform to accomplish a similar result.

In cases like these, one must establish two intermediate coordinate systems. The first one, S for spacecraft, has a known orientation in inertial space. The second one, call it P for "scan platform," carries the camera. Two rotations suffice to transform from S to P, corresponding to the two degrees of freedom inherent in most scan platforms. Another three rotations then specify the orientation of the camera with respect to the movable platform carrying it. These angles are typically small misalignments, and they can often be considered constant during the lifetime of a mission.

Let $\hat{\mathbf{x}}$ represent the desired inertial pointing direction of the camera boresight, often the apparent direction $\hat{\mathbf{A}}$ to the target of interest. Let $\mathbf{S}$ represent the rotation from inertial coordinates to spacecraft coordinates: the orientation of the spacecraft. Similarly, let $\mathbf{M}$ denote the misalignment matrix. A simple representation of $\mathbf{M}$ is

$$\mathbf{M} = \mathbf{R}_3(\theta_3)\ \mathbf{R}_2(\theta_2)\ \mathbf{R}_1(\theta_1), \tag{4.34}$$

where the three θ_i are the misalignment angles, nominally zero. Then the camera orientation could be given by

$$\mathbf{C} = \mathbf{M}\ \mathbf{R}_2(e)\ \mathbf{R}_3(a)\ \mathbf{S}. \tag{4.35}$$

Here a and e are traditionally known as the *azimuth* and *elevation* angles of the scan platform. The third row of **C** contains the transpose or $\hat{\mathbf{x}}$.

To determine a and e from Eq. (4.35), given **S**, **M**, and the pointing direction $\hat{\mathbf{x}}$, is fairly straightforward. We know that $\mathbf{C}\hat{\mathbf{x}} = \hat{\mathbf{k}} = (0,0,1)^{\mathrm{T}}$ by definition. If we multiply both sides on the left by $\mathbf{M}^{\mathrm{T}}$ and define a vector $\hat{\mathbf{s}} \equiv \mathbf{S}\,\hat{\mathbf{x}}$, we obtain

$$\mathbf{R}_2(e)\,\mathbf{R}_3(a)\hat{\mathbf{s}} = \mathbf{M}^{\mathrm{T}}\begin{pmatrix}0\\0\\1\end{pmatrix} \tag{4.36}$$

which, written out, becomes

$$\begin{aligned} \cos e \cos a s_1 + \cos e \sin a s_2 - \sin e s_3 &= M_{31}, \\ -\sin a s_1 + \cos a s_2 &= M_{32}, \\ \sin e \cos a s_1 + \sin e \sin a s_2 + \cos e s_3 &= M_{33}. \end{aligned} \tag{4.37}$$

The second equation is in the form

$$a\cos\theta + b\sin\theta = c,$$

whose solution is found by

$$\begin{aligned} r &= \sqrt{a^2 + b^2}, \\ \theta &= \tan^{-1}(b/a) + \tan^{-1}(c/r). \end{aligned}$$

The first and third of Eq. (4.37) can be combined to eliminate the azimuth, and the solution proceeds similarly.

4.5.3 Primary and Secondary Axes

Some spacecraft, notably Cassini, define a desired attitude in terms of two spacecraft-fixed vectors and two directions in space. A "primary" spacecraft-fixed vector—for instance, the boresight of a camera—is commanded to point at a specific direction or at a specific target. This primary specification eliminates two of the three degrees of freedom in the spacecraft attitude. The third degree of freedom, namely, the roll angle about the primary axis, is eliminated by requiring the secondary axis of the spacecraft and the secondary direction in space to lie in the same plane. This secondary requirement fixes the spacecraft's roll angle about the primary axis. The details are mission-specific, but the algorithm must work something like this:

1. The desired pointing direction of a camera, expressed as a unit vector $\hat{\mathbf{p}}$ in inertial coordinates, fills the third row of **C**.
2. Take the cross product of the boresight and the secondary axis $\mathbf{s}$ in space. Unitize this vector and call it $\hat{\mathbf{x}}$.

3. Take the cross product $\hat{\mathbf{p}} \times \hat{\mathbf{x}}$ and unitize it to yield $\hat{\mathbf{y}}$. By construction, $\hat{\mathbf{y}}$ is perpendicular to the boresight, and the secondary axis in space lies in the plane containing $\hat{\mathbf{p}}$ and $\mathbf{S}$.
4. Depending on the orientation of the spacecraft secondary axis with respect to the camera, one further rotation may be required to determine the position angle of the spacecraft secondary axis in camera coordinates.

4.5.4 Rotation into Camera Coordinates

Regardless of the method by which the camera orientation matrix **C** is calculated, the apparent vector **A** from the camera to any target (body center, limb point, terminator point, landmark, star) must be rotated into camera body coordinates. Historically the resulting vector—from the camera to the target, expressed in camera coordinates— has been denoted by **P**. The transformation is simple:

$$\mathbf{P} = \mathbf{C}\,\mathbf{A}. \tag{4.38}$$

4.6 Modeling the Gnomonic Projection

The mathematics of the gnomonic projection is straightforward. The camera coordinate system C, as depicted in Figure 3.1, has its origin at the pinhole. The third axis L points along the "boresight" of the camera, toward the scene being imaged. The M- and N-axes (first and second, respectively) complete the right-handed triad. It is this coordinate system into which the vector **P** from the camera to any target must be rotated, using the "**C** matrix" discussed in the previous section.

The focal plane of the camera is in effect the back wall of a camera obscura. This plane has its own (x, y) coordinate system. The origin is defined to be the point where **P**, projected backward, intersects the focal plane. Because the scene is projected upside-down, it is convenient to make the x- and y-axes antiparallel to M and N. Therefore the vector (x, y, f) shares the same direction as **P**, and their components have the same ratios:

$$\frac{x}{P_1} = \frac{y}{P_2} = \frac{f}{P_3}. \tag{4.39}$$

This yields the expressions for the ideal image coordinates under gnomonic projection:

$$x = f(P_1/P_3) \text{ and} \tag{4.40}$$

$$y = f(P_2/P_3). \tag{4.41}$$

Figure 3.1 shows a pinhole camera, with the coordinate axes of the camera and those of the image plane included. The image plane, where the detector is located, is the left wall, and the aperture is in the center of the right wall. For an infinitesimally small aperture, and considering only geometric optics, the distance f from the pinhole *is* the focal length of the camera.

Real cameras of course have not pinholes but optical elements; nevertheless, the basis of the mathematics remains the gnomonic projection. The next few sections deal with deviations from it.

4.7 Modeling Distortions and Misalignments

The causes of the various optical aberrations are described in Section 3.2. This section provides two ways of modeling the effects of these aberrations on the location of images within a picture. The model used at JPL was developed by Chuck Acton and Tom Duxbury in the late 1960s, when optical navigation was first demonstrated on Mariner 9 (1971) and the Viking orbiters (1976). A second model, in common use in the machine vision field, is OpenCV (Bradski 2000; Itseez 2015).

4.7.1 The Acton–Duxbury Model

Distortion is the only classical third-order aberration to affect the position of images. It moves each image with respect to the "principal point" by an amount proportional to the cube of the angle between the incoming light and the optical axis or, for practical purposes, to the cube of the distance in the focal plane between the optical axis and the image. Even this is an approximation, and if the distortion is severe enough, one must add a fifth-order term. If, as in the previous section, (x, y) denote the ideal image location, the distorted image location (x', y') can be modeled as

$$\begin{pmatrix} x' \\ y' \end{pmatrix} = \begin{pmatrix} x \\ y \end{pmatrix} + \alpha_2 \begin{pmatrix} xr^2 \\ yr^2 \end{pmatrix} + \alpha_4 \begin{pmatrix} xr^4 \\ yr^4 \end{pmatrix}, \tag{4.42}$$

where the α_i are coefficients giving the magnitude of the distortion and $r^2 \equiv x^2 + y^2$. Most cameras have a measurable α_2, but few have a significant α_4.

Keystone or "tip/tilt" distortion arises if the detector is not strictly perpendicular to the optical axis. If the detector is mounted incorrectly, at an angle to the correct plane, one side of the detector will lie closer to the aperture than the other. Images on the nearer side will therefore appear closer together, while those on the farther side will be spread apart. This phenomenon is a natural consequence of the gnomonic projection.

The model for keystone distortion contains two terms, one for a rotation of the detector about the x-axis, the other for rotation about the y-axis. We have

$$\begin{pmatrix} x' \\ y' \end{pmatrix} = \begin{pmatrix} x \\ y \end{pmatrix} + \alpha_5 \begin{pmatrix} xy \\ y^2 \end{pmatrix} + \alpha_6 \begin{pmatrix} x^2 \\ xy \end{pmatrix}. \tag{4.43}$$

Tangential distortion is a rare effect that rotates images about the optical axis by an *even* power of r. The Acton–Duxbury model includes this effect by

$$\begin{pmatrix} x' \\ y' \end{pmatrix} = \begin{pmatrix} x \\ y \end{pmatrix} + \alpha_1 \begin{pmatrix} -yr \\ xr \end{pmatrix} + \alpha_2 \begin{pmatrix} -yr^3 \\ xr^3 \end{pmatrix}. \tag{4.44}$$

The radial, tangential, and tip/tilt terms are small enough that they can be considered to be independent, and their effects simply add. The full Acton–Duxbury model for converting the ideal image location (x, y) to the observed (x', y') is thus

$$\begin{pmatrix} x' \\ y' \end{pmatrix} = \begin{pmatrix} x \\ y \end{pmatrix} + \begin{pmatrix} -\alpha_1 yr + \alpha_2 xr^2 - \alpha_3 yr^3 + \alpha_4 xr^4 + \alpha_5 xy + \alpha_6 x^2 \\ \alpha_1 xr + \alpha_2 yr^2 + \alpha_3 xr^3 + \alpha_4 yr^4 + \alpha_5 y^2 + \alpha_6 xy \end{pmatrix}. \tag{4.45}$$

4.7.2 The OpenCV Distortion Model

The open source software package *OpenCV* (Bradski 2000, Itseez 2015) uses a distortion model that is similar to but somewhat more comprehensive than the Acton–Duxbury model described above. In this formulation,

$$\begin{pmatrix} x' \\ y' \end{pmatrix} = \begin{pmatrix} x \\ y \end{pmatrix} + \frac{k_1 r^2 + k_2 r^4 + k_3 r^6}{k_4 r^2 + k_5 r^4 + k_6 r^6} \begin{pmatrix} x \\ y \end{pmatrix} + \begin{pmatrix} 2p_1 xy + p_2(r^2 + 2x^2) \\ p_1(r^2 + 2y^2) + 2p_2 xy \end{pmatrix}, \tag{4.46}$$

where the k_i and p_i are constants to be determined by calibration.

This model has distinct distortion parameters in x and y, unlike the Acton–Duxbury model, and therefore it can accommodate separate radial distortion calibrations in x and y. The tip/tilt terms (coefficients α_5 and α_6) are accounted for in OpenCV by coefficients p_1 and p_2, if one can subsume their r^2 contributions into k_1. The terms in the denominator (coefficients k_4 through k_6) are not present at all in the Acton–Duxbury model.

4.8 Conversion into Pixel Coordinates

The previous sections have dealt with optics: the gnomonic projection of a scene onto the image plane, as well as optical distortions that give rise to deviations from the ideal projection. This section describes the final step in the creation of an picture: the mapping from (x, y) coordinates in the image plane into the pixels that comprise an actual digital image.

Photographic emulsions, whether deposited onto glass plates or onto plastic film, consist (after development) of myriads of irregularly sized, irregularly spaced "grains." It is altogether appropriate to continue to use rectilinear coordinates (x, y) to describe the location of images on such a picture.

Digital pictures, however, are formed differently. The mechanisms by which a digital image is created are discussed in Chapter 3, but for the purposes of this section it will suffice to say that a *pixel* is nominally a square or rectangular *area* in the image plane and that a picture comprises a rectangular array of such pixels. A digital picture thus consists of a rectangular array of numbers. In JPL parlance, these numbers are *data numbers* ("DN"). (The term "analog-to-digital unit" or ADU is also found in the literature.)

A linear transformation from focal plane coordinates into pixels is:

$$\begin{pmatrix} s \\ l \end{pmatrix} = \begin{pmatrix} s_0 \\ l_0 \end{pmatrix} + \begin{pmatrix} K_x & K_{xy} \\ K_{yx} & K_y \end{pmatrix} \begin{pmatrix} x \\ y \end{pmatrix}, \tag{4.47}$$

where (s_0, l_0) are the pixel coordinates of the optical axis—that is, where the optical axis appears in the digital picture—and the so-called "**K** matrix" (dimensioned pixels/mm) provides not only scale factors in each axis but also a rotation. This formulation allows pixels to be parallelograms.

It must be noted here that some cameras produce a mirror image of the sky. This phenomenon arises if either (i) there is an odd number of mirrors in the optics or (ii) the detector is backside-illuminated. Sometimes, despite a careful reading of the camera documentation, this feature is not noticed until the first calibration frames are taken after launch. The remedy for this situation is to set K_y to a negative value, in effect flipping the picture about the horizontal axis.

JPL's software adds a third column to **K**, so that

$$\begin{pmatrix} s \\ l \end{pmatrix} = \begin{pmatrix} s_0 \\ l_0 \end{pmatrix} + \begin{pmatrix} K_x & K_{xy} & K_{xxy} \\ K_{yx} & K_y & K_{yxy} \end{pmatrix} \begin{pmatrix} x \\ y \\ xy \end{pmatrix}. \tag{4.48}$$

This quadratic term seems to have been necessary in order to model the Voyager television cameras properly. Few modern detectors have required it.

The decision as to whether the first pixel in a picture should be called (0,0) or (1,1) is arbitrary, as is the decision as to whether the integer values $(s.0, l.0)$ should be in the center of a pixel or in a corner.

4.9 Summary of Optical Navigation Geometry Calculations

1. Obtain the position $\mathbf{R}(t)$ and velocity $\dot{\mathbf{R}}(t)$ of the camera with respect to the Solar System barycenter at the time t corresponding to the midpoint of the exposure.
2. Obtain the rotation matrix **C** which gives the attitude of the camera with respect to inertial coordinates at the midpoint of the exposure.

3. For each object:
   a. Obtain iteratively the true position $\mathbf{T}(t)$ of the object with respect to the camera:

$$\mathbf{T}(t) = \mathbf{S}(t - \tau) - \mathbf{R}(t),$$

   where $\tau = |\mathbf{R}(t) - \mathbf{S}(t - \tau)|/c$.

   b. Obtain the apparent position

$$\mathbf{A}(t) = \mathbf{T}(t) + \tau\dot{\mathbf{R}}(t).$$

4. For each star:
   a. Apply proper motion (and radial velocity if necessary) to propagate the star's right ascension and declination to the midpoint of the exposure. Use the star's parallax, if known, to get the position $\mathbf{S}(t)$ of the star relative to the Solar System barycenter.
   b. Calculate the true position:

$$\mathbf{T}(t) = \mathbf{S}(t) - \mathbf{R}(t).$$

   c. Calculate the apparent position:

$$\hat{\mathbf{r}} = \mathbf{T}(t)/|\mathbf{T}(t)|;$$

$$\mathbf{A}(t) = \mathbf{r} + \dot{\mathbf{R}}(t)/c.$$

5. Rotate the apparent position into camera coordinates:

$$\mathbf{P} = \mathsf{C}\,\mathbf{A}.$$

6. Use the gnomonic projection to obtain the image coordinates in the image plane:

$$\begin{pmatrix} x \\ y \end{pmatrix} = \frac{f}{P_3}\begin{pmatrix} P_1 \\ P_2 \end{pmatrix}.$$

7. Account for optical aberrations, misalignments, and other distortions as necessary.
8. Convert into pixel coordinates:

$$\begin{pmatrix} s \\ l \end{pmatrix} = \begin{pmatrix} s_0 \\ l_0 \end{pmatrix} + \begin{pmatrix} K_x & K_{xy} & K_{xxy} \\ K_{yx} & K_y & K_{yxy} \end{pmatrix}\begin{pmatrix} x \\ y \\ xy \end{pmatrix}.$$

5

Obtaining Optical Navigation Observables

5.1 Introduction

This chapter is devoted to *image processing*, the task of extracting the optical navigation *observables* from a picture. We assume that the characteristics of the camera are known and that the picture has already been flattened and any readout smear removed, so that the picture is as close as one can get to a faithful rendition of the scene being imaged. We further assume that the noise in each pixel has been calculated.

Sections 5.2–5.6 describe various algorithms for inferring the center of images of stars and unresolved bodies, of larger bodies, and of features on bodies, for the choice of centerfinding technique can depend on the size and shape of the body being imaged. This is still an area of active research, and the algorithms described here are by no means a complete set. It has become routine practice in flight to apply several different techniques to the same pictures, and then to examine the postfit residuals (supplied by the greater navigation team) to determine which technique has done the best job of fitting the data.

5.2 Centerfinding for Stars

Stars are, without exception, so distant from the camera that they may be considered point sources. The *image* of a star within a picture, however, is not a two-dimensional Dirac delta function. Rather, owing to diffraction within the camera optics, a star image is broadened into a *point-spread function*, universally abbreviated to "PSF." This phenomenon is beneficial for optical navigation. If a star image were a delta function, its signal would be entirely contained within one pixel, and it would be impossible to obtain accurate pixel coordinates for the image. One would be forced to place the observed image coordinates at the center

Spacecraft Optical Navigation, First Edition. William M. Owen Jr.

of the pixel, with an uncertainty of half a pixel. Note that this uncertainty would not be a Gaussian sigma, as the underlying distribution is uniform rather than Gaussian.

Because a camera's PSF is typically broader than one pixel, starlight usually illuminates more than one pixel. Dim star images, well centered in a pixel, may encompass only that one pixel, but in most cases a star image will contain signal in two to nine pixels, and exceptionally bright stars may produce images that are tens of pixels across.

Many methods for star centerfinding model the PSF as some circularly symmetrical function with a peak at the origin. Suppose the modeled star image is centered at sample s_c and line l_c. (The subscript c denotes "center." Do not confuse these with s_0 and l_0, the coordinates of the optical axis of the camera.) The intensity I at some arbitrary (s, l) would then be given by some arbitrary function

$$I = f(r) \tag{5.1}$$

with

$$r^2 = (s - s_c)^2 + (l - l_c)^2. \tag{5.2}$$

The predicted brightness of the image within a rectangular area of the focal plane—for example, within a pixel—is found by integrating the intensity:

$$B = \int_{l_1}^{l_2} \int_{s_1}^{s_2} I(s, l)\, ds\, dl, \tag{5.3}$$

where the rectangle is defined by $s_1 \leq s \leq s_2$ and $l_1 \leq l \leq l_2$. As we need to fit this model to the actual picture, we must develop the partial derivatives of B not only with respect to (s_c, l_c) but also with respect to the other parameters of the fitting function: typically a height, a width, and a constant background. The fit uses a least-squares approach, in which the observed quantities are the data number (DN) in those pixels under consideration (typically a rectangular region centered on the predicted image location). As the problem is inherently nonlinear, the solution is iterated until convergence is achieved.

5.2.1 Circular Gaussian Model

Perhaps the simplest approach is to model the star's intensity as a Gaussian:

$$I = \frac{h}{\sqrt{2\pi\sigma^2}} e^{-r^2/2\pi\sigma^2}. \tag{5.4}$$

Although the Gaussian function is not analytically integrable, it is easy enough to expand the exponential into a Taylor series, which one can integrate to yield the expected DN in a pixel. One can substitute the definition for r^2 in Eq. (5.2) and thereby split the expression for I into a product of two exponentials,

$$I = he^{-(s-s_0)^2/2\pi\sigma^2}\ e^{-(l-l_0)^2/2\pi\sigma^2},$$

and then the brightness within the pixel located at (p, l) is found by

$$B = b + h \int_{l_1}^{l_2} e^{-(s-s_0)^2/2\pi\sigma^2}\ dl \int_{s_1}^{s_2} e^{-(l-l_0)^2/2\pi\sigma^2}\ ds$$

with b a constant background. Each of the two integrals may be evaluated by expanding the exponential into a Taylor series and then integrating the series term by term.

5.2.2 Elliptical Gaussian Model

If the axes of the model ellipse are restricted to be parallel to the sample and line axes of the detector, then one can replace σ in Eq. (5.4) by either σ_s or σ_l as appropriate.

5.2.3 Circular Cauchy Function

Another model, also circularly symmetric with a peak at the origin, is the Cauchy or Lorentz distribution,

$$I = \frac{h\sigma^2}{\sigma^2 + r^2},$$

where I, h, and r are defined as above, and σ is the half-width at half-maximum of the distribution. This function does not seem to have an easy analytic integral. Although centerfinding for the Voyager mission used this model, a closer inspection of Riedel *et al.* (1990) shows that the PSF was never integrated over the area of a pixel. Rather, the star images were usually trailed due to camera attitude motion during the exposure, and the integration was performed over the length of the "smear vector" and evaluated only at the center of each pixel. The software solved for the length and direction of the smear vector as well as for the pixel coordinates of the star image.

5.2.4 Centerfinding Using the Marginal Distribution

One method that does not rely on a model for the PSF of the camera uses the *marginal distribution* of the DN array. Let (s_m, l_m) be the coordinates of that pixel within the star image which has the maximum DN value. Presumably the center of the star image will lie close to this pixel.

Extract a 3×3 block of pixels centered at (s_m, l_m). The sample coordinate of the star is found as follows. For each column, add the DN in the three extracted pixels in that column to get the sum. Call these B_-, B_0, and B_+ for columns $(s-1)$, s, and $(s+1)$, respectively. In most cases the three points $(s-1, B_-)$, (s, B_0), and $(s+1, B_+)$ define a parabola, opening downward,

$$B(s) = as^2 + bs + c, \tag{5.5}$$

with $a < 0$. We can solve for a and b (c is not needed) by

$$2a = B_+ + B_- - 2B_0 \tag{5.6}$$

and

$$b = (B_+ - B_-)/2. \tag{5.7}$$

The vertex of the parabola becomes the sample coordinate of the star:

$$s_c = s_m - b/2a. \tag{5.8}$$

The uncertainty in the answer may be found in two different ways, either by propagating the variances of the DN array, giving

$$\sigma^2(s_c) = \left[(B_0 - B_-)^2\sigma^2(B_+) + (B_+ - B_0)^2\sigma^2(B_-) + (B_+ - B_-)^2\sigma^2(B_0)\right] / 4a^2, \tag{5.9}$$

or more simply from the width of the fitted parabola:

$$\sigma^2(s_c) = |(b^2 - 2as_c)/4a|. \tag{5.10}$$

The line coordinate and its uncertainty are found similarly, by adding the DN values across the three rows of the extracted block.

This method will fail in one coordinate or the other if $B_- - B_0 = B_+$, for then $a = 0$ and $-b/2a$ is undefined. If B_0 is the largest of the three B's, the solution will lie within half a pixel of s_m and l_m. If B_0 is equal to either B_- or B_+, the solution will be exactly 0.5 pixel less than or greater than either s_m or l_m, as the case may be.

5.3 Reflectance Laws

To find the center of the image of a Solar System body—that is, the point in the picture to which the physical center of the body would project—usually requires some knowledge of how incident sunlight is reflected or scattered, particularly in the direction toward the camera. The reflectance law must therefore be used to predict how much light is expected to fall in each pixel of an image of that body. Nevertheless, Minnaert (1941) writes, "The surface brightness of a solid body, illuminated by parallel rays from a definite direction and observed from another direction, cannot be expressed by a simple physical law; the formulae thus far proposed are rather empirical and vary from one substance to another." It is for precisely this reason that different reflectance laws exist.

The geometry behind a reflectance law is not hard to grasp. At any point on the surface of a body there exists the *surface normal*, a unit vector $\hat{\mathbf{n}}$ perpendicular to the surface at that point. (We assume that the surface has no sharp edges!) Sunlight impinges on the surface, and the unit vector from the surface to the sun will be called $\hat{\mathbf{s}}$. Finally, there is a unit vector $\hat{\mathbf{c}}$ pointing from the surface toward the camera. Figure 5.1 is a diagram showing these vectors.

These three vectors define three angles:

1. The *angle of incidence* i is the angle between the incident sunlight and the normal to the surface. For *normal incidence*, when the sun is "directly overhead," $i = 0$; for *grazing incidence*, when the sun is approaching the local horizon, $i \to 90°$. It is evident that $i = \cos^{-1}(\hat{\mathbf{n}} \cdot \hat{\mathbf{s}})$ and that i can never exceed 90°.
2. The *angle of emission* e is the angle between the outgoing or "scattered" light and the surface normal. As with i above, e is defined as $e = \cos^{-1}(\hat{\mathbf{n}} \cdot \hat{\mathbf{c}})$, and it too cannot exceed 90°.
3. The *phase angle* ϕ is the angle between the incident sunlight and the scattered light, $\phi = \cos^{-1}(\hat{\mathbf{s}} \cdot \hat{\mathbf{c}})$. Unlike i and e, ϕ can take on values between 0° and 180°.

Note that the term "phase angle" can also be applied to the entire body. This usage, indeed the term "phase" itself, hearkens to the phases of the moon: a full

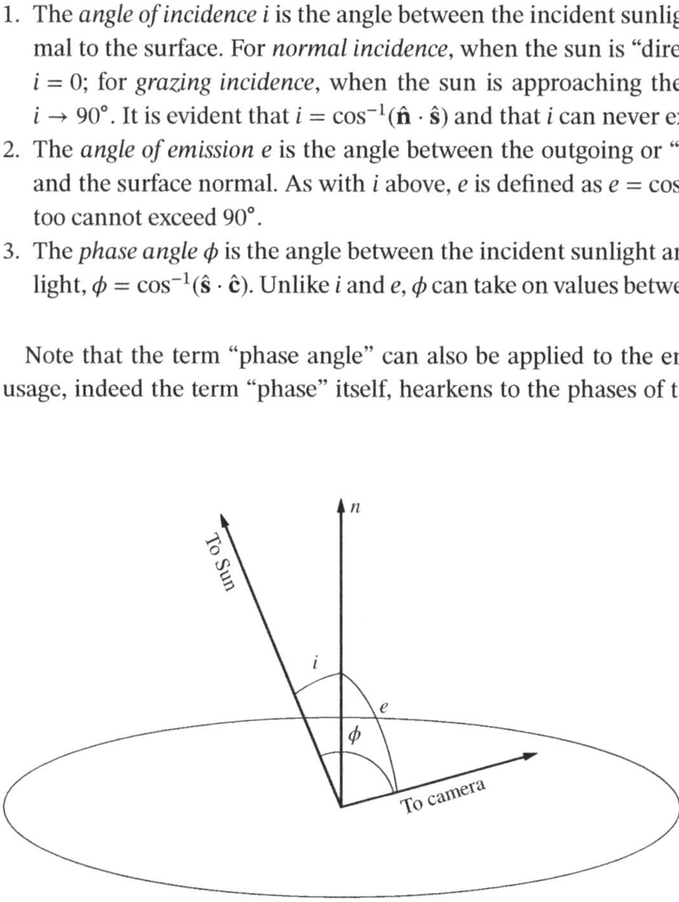

Figure 5.1 Incidence, emission, and phase angles, with vectors pointing toward the sun, the camera, and the normal to the surface.

moon would have a very small value of ϕ, first and last quarter would have ϕ close to 90°, and new moon would have ϕ nearly 180°.[1]

A reflectance law, also called a "scattering law" because it describes how sunlight is "scattered" into 2π steradians, describes how incident light is re-emitted from a surface, as a function of all three angles. The general model is

$$I = AI_0 f(i, e, \phi)/r^2 \tag{5.11}$$

where I_0 is the intensity of sunlight, r is the distance of the body from the sun, and A is the body's *albedo*, which specifies how much of the incident sunlight is reflected. For opnav purposes, the *absolute* value of I is not needed. It is far more important to model *changes* in I over the image. Several reflectance laws are in common use.

5.3.1 Lambert Scattering

The simplest and oldest reflectance model was developed by Johann Heinrich Lambert in 1760: incident light is reflected isotropically. Therefore the intensity I of reflected light is proportional only to the cosine of the angle of incidence, without regard to the angle of emission. We thus have

$$I \propto \cos i. \tag{5.12}$$

5.3.2 Lommel–Seeliger Scattering

In this model, developed originally by Eugen von Lommel (1887) and refined by Hugo von Seeliger based on observations of the moon,

$$I \propto \frac{\cos i}{\cos i + \cos e}. \tag{5.13}$$

There is still a factor $\cos i$ in the numerator, as there must be, but the presence of e indicates that the scattering is no longer isotropic. Compared to Lambert, there is less light emitted at low e (that is, upward relative to the surface) and more at high e (more parallel to the surface). This model is particularly good for the moon and Mars, rocky bodies with significant regolith.

1 New moon, first quarter, full moon, and last quarter are not defined by phase angle. Rather, they are defined as the moments when the apparent longitudes of the sun and moon differ by multiples of 90° as seen from the center of the earth. The moon has $\phi = 90°$ at *dichotomy*, when the right angle in the sun–earth–moon triangle is at the moon, and historically the angular difference between first or last quarter and dichotomy has been used to estimate the ratio of the distances of the sun and moon from the earth.

5.3.3 Minnaert Scattering

Minnaert (1941), having undertaken extensive photometric observations of the moon, proposed the reflectance model

$$I \propto (\cos i)^k \; (\cos e)^{k-1}, \tag{5.14}$$

where the parameter k controls the relative importance of the two cosines. If $k = 1$ the expression reduces to Lambert scattering.

5.3.4 Hapke Scattering

Bruce Hapke, in a series of papers between 1981 and 2002, developed perhaps the most realistic model of scattering from the regolith of an airless, rocky body. His model accounts for multiple reflections, the *opposition surge* that arises from the lack of shadows at $i = 0$, and the tendency of scattered light to remain in the plane containing the sun and the zenith, known as *forward scattering.*

Hapke's model has the form

$$I \propto \frac{\cos i}{\cos i + \cos e} \left[p(\phi)B_{\mathrm{SH}}(\phi) + M(\cos i, \cos e)\right] B_{\mathrm{CB}}(\phi), \tag{5.15}$$

where the functions $p(\phi)$, $B_{\mathrm{SH}}(\phi)$, $M(\cos i, \cos e)$, and $B_{\mathrm{CB}}(\phi)$ are described in Hapke (2002). The values of the variables found in these functions are determined from analyzing resolved images of the body in question, taken at a variety of phase angles. It is evident that the leading term represents Lommel–Seeliger scattering, with the remaining term being a correction to that model.

5.4 Centerfinding for Unresolved Bodies

If a body is so small or so distant that its apparent diameter is smaller than the camera's PSF, the same methods that are used for star images can be used to find the *photocenter* or "center of light" of the body's image. Even though the offset of the photocenter from the physical center of the body must be small in terms of pixels, it is systematic. Unless the shift is modeled and corrected, the resulting navigation solution will contain a systematic error.

The photocenter offset vector is easily defined in pixel coordinates by

$$\Delta s = \int s\Phi(s,l)\, ds\, dl \Big/ \int \Phi(s,l)\, ds\, dl \text{ and} \tag{5.16}$$

$$\Delta l = \int l\Phi(s,l)\, ds\, dl \Big/ \int \Phi(s,l)\, ds\, dl, \tag{5.17}$$

where $\Phi(s, l)$ is the reflectance function of the object evaluated at those image coordinates. This function actually depends on the body's albedo, the angle of incidence i at the surface, the angle of emission e, and the phase angle ϕ. In general this integration must be performed numerically.

If the body is a sphere and Lambert's scattering law holds, both integrals can be done analytically. The result for the *magnitude* R of the photocenter offset is

$$R = \frac{3r \sin\phi(1 + \cos\phi)}{16[\sin\phi + (\pi - \phi)\cos\phi]} \tag{5.18}$$

where r is the radius of the spherical body. One must then transform R (a scalar, measured in units of length) into offsets in sample and line coordinates. One way to accomplish this task—certainly not the only way—is to set up a coordinate system in which the z-axis points from the camera to the body and the x-axis points from the body to the sun. In this coordinate system the offset vector is $(R, 0,0)$. One then rotates this vector back into inertial coordinates, adds it to the known vector from the camera to the center of the body, and transforms the results into (s, l) in the usual way. The difference between these pixel coordinates and those of the body center is the correction that must be applied to the *predicted* coordinates of the body center before the residual is found.

5.5 Centerfinding for Resolved Bodies

If a body is so large or so near that its apparent angular diameter is significantly greater than the PSF of the camera, its image is said to be *resolved*. Centerfinding methods that are appropriate for point sources are no longer applicable. Two methods have been in common use:

1. Model the expected brightness of each pixel in the entire body and then correlate that model against the actual picture to determine the center.
2. Find points on the image of the limb and possibly of the terminator of the object and then use those points to infer the center.

5.5.1 Correlation

Given the geometry associated with a picture—the vectors giving the position of the sun and of the camera relative to the body, in body-fixed coordinates—the parameters describing the camera, and an appropriate reflectance law, one can calculate a simulated or predicted image of the body, built up pixel by pixel. The pixel coordinates of the center of the body in the simulated image are known by construction. Let the pixel coordinates of the simulated image be numbered such that (0,0) represents the center of the body, and suppose that the simulated image has limits $(-N_s, +N_s)$ in the sample direction and $(-N_l, +N_l)$ in the line direction. This implies that the simulated image has an odd number of samples and lines, a condition that while not strictly necessary is quite convenient in practice.

The correlation $C(s, l)$ is the product, pixel by pixel, of the simulated image S and the actual picture P. It is primarily a function of the location (s, l) in the picture of where the center of the simulated image is placed, although of course it is also a function of the fidelity of the simulated image. The *correlation score* for a given shift is defined by

$$C(s,l) = \sum_{i=-N_s}^{+N_s} \sum_{j=-N_l}^{+N_l} P(s+i, l+j)S(i,j). \tag{5.19}$$

If the method is successful, there will be a *correlation peak*, usually one global maximum of C but occasionally two or four neighboring pixels will share the same score.

In most cases shifting the offset by one or two pixels away from the "best" offset will still produce a significant correlation. If so, the correlation array behaves very much like a pixelated star image, and any of the methods used for centroiding stars can be applied. Jet Propulsion Laboratory (JPL) software has used the marginal distributions, as described in Section 5.2.4, to interpolate the correlation array and locate its maximum.

Correlation can be tricky even with modern hardware. Albedo variations and shadowing from topographical features (especially near the terminator), if not correctly modeled, can overwhelm the information content near the limb and produce erroneous results. Restricting the area of correlation to only those regions within a few pixels of the limb or terminator may be a viable option, but this author knows of no implementation of that technique.

5.5.2 Limb Scanning

In the Voyager and Galileo eras, computer memory and storage capabilities were not large enough to enable whole-body correlation, as the necessary arrays would have been several hundred pixels on each side. The preferred method—indeed, the only practical method at the time—was "limb scanning" (Breckenridge *et al.* 1972, Campbell *et al.* 1980, Campbell *et al.* 1982, Riedel *et al.* 1990). This method is a two-step process.

First, one constructs a set of radial *scan lines* emanating from the predicted center of the body. In three dimensions, these are actually half-planes bounded by the line containing the center of the body and the camera. Each plane intersects the body in some curve on the body's surface. If the body can be modeled as a sphere, that curve is a diameter of the sphere. If the body is a triaxial ellipsoid, the curve is an ellipse. If the body's shape is so irregular that it must be modeled as a collection of triangular plates, the scan planes will intersect the shape model in a set of connected line segments. Regardless of the model for the body, the intersection of the surface and a plane can be found.

Define the *scan angle* ψ as the angle *in the picture* defined by the $+x$-axis and the scan line; the vertex lies at the image coordinates of the predicted center of the body. Let (S_i, L_i) be the image coordinates of the predicted limb or terminator point that lies on the ith scan line.

One proceeds to model the expected brightness, pixel by pixel, along each scan line. Historically the model has assumed a constant albedo, relying only on variations in the angles of incidence and reflection to compute the predicted brightness levels. There is no reason, however, why albedo variations and topographic variations could not be included in the model. The camera's point spread can be accounted for as well, by considering how neighboring terrain will contribute to the brightness levels of the pixel in question. The result of these calculations is a set of one-dimensional predicted values, one set for each scan line, with $(2P + 1)$ predictions in each scan line.

The observed brightness along each scan line is found by interpolating the DN values in the relevant pixels. Historically this has been done by bilinear interpolation using the DN values of the pixel in question and its three nearest neighbors. This technique tacitly assumes that the DN values are the values of a continuous function *evaluated at the center of a pixel*, rather than as the integral of a brightness distribution. This assumption is of course not correct, but the computers used in the Voyager era simply could not support anything more rigorous. Each scan line thus contains $(2N + 1)$ interpolated DN values d_i spaced at intervals of one pixel, with d_0 centered at the computed limb.

These two one-dimensional brightness arrays for each scan—the interpolated array d_i from the picture and the computed array p_i from the reflectance law—are then correlated to give an array c_i with $(2C + 1)$ entries, where $C = N - P$:

$$c_i = \sum_{j=1}^{2P+1} p_j d_{i+j}. \tag{5.20}$$

The maximum value of the c_i is found, and then the method described in Section 5.3 (or some other) is used to determine the subpixel location λ of the correlation peak for that scan line. Clearly $-C < \lambda < +C$ for a usable scan, and $\lambda = 0$ if the model is perfect and the assumed center of the body image is correct. The scan result can be transformed into pixel coordinates:

$$s_i = S_i + \lambda \cos \psi_i; \tag{5.21a}$$

$$l_i = L_i + \lambda \sin \psi_i. \tag{5.21b}$$

The result is a set of measured pixel coordinates for limb or terminator points, one point along each scan line.

Second, these measured points must be used to infer the center of the body. A least-squares fit is used, with the λ_i (the difference between the observed and computed limb or terminator points) as the prefit residuals. The partial derivatives for the limb of a spherical body depend only on the scan angle:

$$\partial\lambda_i/\partial s = \cos\psi_i; \tag{5.22a}$$

$$\partial\lambda_i/\partial l = \sin\psi_i; \tag{5.22b}$$

The foregoing treatment assumes that the body is spherical or nearly so. If the body is noticeably a triaxial ellipsoid—for instance, Jupiter's four Galilean satellites and Saturn's larger satellites (but not Titan)—one often uses scans along a line normal to the limb instead of a truly radial scan. In this case the scan angle ψ retains its definition in the construction of the predicted and observed scans, but there is now a different angle, the "search angle," which is measured from the presumed center of the image. This author's advice in such situations is to set up parallel runs and see which method produces better residuals, for there are not necessarily "right" and "wrong" methods, but rather effective and ineffective ones.

5.6 Centerfinding for Landmarks

A landmark is defined mathematically by a vector, constant in body-fixed coordinates, emanating from the center of a body; the tip of the vector *is* the landmark. In practice, the tip of the vector is defined relative to features on the surface: the center of a crater, for instance, or perhaps a distinctive point on a rock. "Natural" landmarks like these are amenable to processing by the human eye alone. An analyst locates the image of the landmark within a picture and notes its sample and line coordinates. The result is reliable only to the extent that the human eye/brain combination can reliably find the same point regardless of lighting conditions or foreshortening (as evidenced by the angles of incidence and reflection, respectively). Experience with the NEAR–Shoemaker mission to Eros (Owen *et al.* 2001) indicated that this primitive method could actually be reliable to about one pixel, if the feature was small and well defined.

Robert Gaskell has spent most of his career, first at JPL and later elsewhere, developing a system that uses *stereophotoclinometry* ("SPC" for short) to define and measure landmarks on the surface of a solid body. Stereo imaging—that is, pictures taken at different angles of emission, providing different views of the same terrain—can provide the *elevation* (height above some reference surface) for salient features. Photoclinometry, or "shape from shading," uses the measured surface brightness to deduce the angles of incidence and emission, assuming a pertinent scattering law. Gaskell's breakthrough (for instance, Gaskell 2011) was to combine the two techniques into a rigorous system.

In Gaskell's method, a landmark includes not only the body-fixed vector itself (which for navigation purposes *defines* the landmark) but also heights and relative albedos for an array of surface points surrounding the vector. (The preferred dimension of the array is 99×99.) In Gaskell's terminology, the heights and relative albedos, taken together, form a *landmark map* or "Lmap." The variation in height from point to point within an Lmap provides the slope of the surface. The slope, combined with the albedo at that point, allows SPC to predict the relative brightness at that point. In this way SPC can create a *predicted* image of the terrain surrounding the landmark. It is then a fairly simple matter to correlate that predicted image against an incoming picture to locate the image of that landmark within the picture.

In order to use SPC to advantage, one must obtain pictures of the surface at a wide variety of incidence and emission angles. This variation is necessary in order to determine, for instance, whether that anomalously dark region is actually darker than its surroundings (lower albedo) or merely on the side of a hill that is receiving less sunlight. Missions that are to orbit asteroids or comets—Dawn, Rosetta and OSIRIS-REx for example—will take "rotation movies," each being a set of pictures taken during a complete rotation of the target. The movies are taken at varying sub-spacecraft latitudes and at varying phase angles, so that all the sunlit terrain can be imaged at a variety of angles of incidence and emission. Usually several such campaigns occur, at ever-decreasing distances, to allow the development of progressively finer Lmaps.

The process of map development begins with an array of overlapping Lmaps covering the entire surface, using (say) a triaxial ellipsoid or a sphere to compute the lengths of the landmark vectors. The Lmaps typically have a center-in-corner overlap pattern, so that any spot on the surface will typically be included in four Lmaps. Initially all the relative heights are zero, and all the albedos are the same. Each incoming picture provides information about the relative albedos and the slopes, to allow an initial estimation of the Lmaps. Next, overlap constraints are supplied, for the same terrain must have the same geography in every Lmap containing it. Enforcing the overlap constraints will also cause the landmark vectors to shift a bit and the Lmaps to come to agreement in the terrain they share. This process also redetermines the observed pixel coordinates of the landmarks, not to mention a better model for the pole and rotation of the target body, information that is passed along to the navigation team so that they can revise their estimate of the spacecraft trajectory.

The above process is iterated as necessary: the updated trajectory is used to produce new (presumably better) estimates of the landmark vectors and body orientation, which produces new opnav measurements, which produces a new trajectory.

It is typical that small-body missions begin with a slow approach, during which the phase angle and sub-spacecraft latitude change very little. The rotation of the body provides enough variation in viewing geometry to enable the creation of Lmaps at progressively finer resolution, but these are only preliminary. Once the spacecraft is in orbit—a much wider variety of viewing conditions is available, and mapmaking begins in earnest. The radiometric data are now sensitive to the body's pole orientation and rotation rate through the irregularities in the body's gravitational field. The optical data are sensitive to the body's orientation too, and the combination of optical and radio data is quite effective, not only in obtaining a better solution for the spacecraft trajectory but also in pointing out possible systematic errors in the landmarks.

6

Using Opnav Data in Orbit Determination

The planning, execution, and analysis of optical navigation imaging are not ends in themselves but only means to an end: the successful navigation of a spacecraft to satisfy mission requirements. Therefore, one must

- calculate the partial derivatives of the optical observations with respect to the various parameters to be estimated in the orbit determination process,
- determine which optical parameters must be estimated and how those should be treated, and
- introduce the optical data to the orbit determination filter properly.

These three topics—the partial derivatives, the construction of the parameter list, and the weighting of the optical data—form most of the material for this chapter.

A fourth topic is camera calibration, the process by which the various camera parameters are determined.

6.1 Dynamic Partials

The dynamic parameters are those which affect the position or velocity, relative to the Solar System barycenter, of the camera or of the object being imaged. "Object" here can mean the center of a body, a landmark on the surface of a body, or a point on the limb or terminator of a body. Some parameters can affect both the camera and the object: for instance, changing the mass (*GM*) of Saturn will change not only the trajectory of a spacecraft in the Saturnian system but also the orbits of the various satellites being imaged. The examples below are intended to be representative, not exhaustive.

The spacecraft trajectory depends upon

- the position and velocity of the spacecraft (if the trajectory has been integrated numerically) at the start time or "epoch" of the integration;

Spacecraft Optical Navigation, First Edition. William M. Owen Jr.

- the gravity of the sun, the planets, and any other nearby objects;
- "solar radiation pressure" resulting from sunlight hitting the spacecraft and being absorbed or reflected;
- trajectory change maneuvers;
- attitude change maneuvers, if accomplished using thrusters;
- atmospheric drag, significant not only for Earth orbiters but also at Mars, Venus, and Titan;
- miscellaneous nongravitational accelerations unique to each spacecraft.

The proper modeling of these forces is outside the scope of this book, as it is the purview of the larger navigation team working together with the rest of the flight team. The team will typically produce, by numerical integration, a trajectory file that also includes partial derivatives; the optical navigation software then queries this file to obtain the spacecraft's position, velocity, and the partial derivatives of these with respect to the parameters of interest.

Similarly, the orbits of the planets and their satellites depend on the parameters of the model used to generate them. If numerical integration is used, as is typical, the parameters include the position and velocity vectors, the mass of the sun, the masses of the bodies being integrated, and (for satellites) the direction of the planet's spin axis, and some of the planet's spherical harmonic coefficients. This list is representative, not exhaustive.

The orbit of an asteroid is also numerically integrated. The integration depends on the initial position and velocity of the asteroid itself and on the masses of the sun and planets. In some cases other asteroids may exert a measurable force on the subject asteroid, in which case their orbital parameters would also be included in the integration. For all natural bodies in the solar system, Jet Propulsion Laboratory (JPL) produces one set of files containing polynomial coefficients for the bodies' position and velocity vectors and a second set of files for the partial derivatives.

In the equations that follow, the letter q is used to denote any of these parameters.

The partial derivatives of the spacecraft's position and velocity with respect to these parameters are $\partial\mathbf{R}(t)/\partial q$ and $\partial\dot{\mathbf{R}}(t)/\partial q$, respectively. These are found by interpolating the appropriate files at the time t of the midpoint of the exposure. As with the position vector itself, the partials are referred to the barycenter of the solar system.

The partials of the observed object's apparent position are likewise given by $\partial\mathbf{S}/\partial q$. The partials are found at the retarded time $(t - \tau)$.

These partials must then be transformed into pixel coordinates. First obtain the partial derivatives of the "true" (unaberrated) vector from the camera to the target by differentiating Eq. (4.2):

$$\frac{\partial\mathbf{T}}{\partial q} = \frac{\partial\mathbf{S}(t-\tau)}{\partial q} - \frac{\partial\mathbf{R}(t)}{\partial q}. \tag{6.1}$$

Note that some of the parameters may affect only **S**, others may affect only **R**, and others may affect both.

Next, stellar aberration must be accounted for. Differentiate Eq. (4.3) to obtain

$$\frac{\partial \mathbf{A}}{\partial q} = \frac{\partial \mathbf{T}}{\partial q} + \tau \frac{\partial \dot{\mathbf{R}}(t)}{\partial q} + \frac{\partial \tau}{\partial q} \dot{\mathbf{R}}(t). \tag{6.2}$$

The partials must then be rotated into camera coordinates. The full expression is

$$\frac{\partial \mathbf{P}}{\partial q} = \mathbf{C}\frac{\partial \mathbf{A}}{\partial q} + \frac{\partial \mathbf{C}}{\partial q}\mathbf{A}, \tag{6.3}$$

although the second term is zero unless **C** has some dependence on the parameter in question—for instance, if one of the principal axes of the camera system depends on the direction from the spacecraft to the sun. (Obviously **C** depends on the pointing angles. Those partials are covered in Section 6.2.)

These partials must be transformed into undistorted millimeters in the focal plane:

$$\begin{pmatrix} \partial x/\partial q \\ \partial y/\partial q \end{pmatrix} = \frac{f}{A_3}\begin{pmatrix} \partial A_1/\partial q \\ \partial A_2/\partial q \end{pmatrix} - \frac{f}{(A_3)^2}\begin{pmatrix} A_1 \\ A_2 \end{pmatrix}\frac{\partial A_3}{\partial q}. \tag{6.4}$$

The camera distortions are accounted for next. As we have in general

$$\begin{pmatrix} x' \\ y' \end{pmatrix} = \begin{pmatrix} x \\ y \end{pmatrix} + \begin{pmatrix} \Delta x \\ \Delta y \end{pmatrix}, \tag{6.5}$$

the partials become

$$\begin{pmatrix} \partial x'/\partial q \\ \partial y'/\partial q \end{pmatrix} = \begin{pmatrix} \partial x/\partial q \\ \partial y/\partial q \end{pmatrix} + \begin{pmatrix} \partial \Delta x/\partial q \\ \partial \Delta y/\partial q \end{pmatrix}. \tag{6.6}$$

However, Δx and Δy are themselves functions of both x and y, and therefore through the chain rule we must have

$$\frac{\partial x'}{\partial q} = \frac{\partial x}{\partial q} + \frac{\partial \Delta x}{\partial x}\frac{\partial x}{\partial q} + \frac{\partial \Delta x}{\partial y}\frac{\partial y}{\partial q} \quad \text{and} \tag{6.7}$$

$$\frac{\partial y'}{\partial q} = \frac{\partial y}{\partial q} + \frac{\partial \Delta y}{\partial x}\frac{\partial x}{\partial q} + \frac{\partial \Delta y}{\partial y}\frac{\partial y}{\partial q}, \tag{6.8}$$

with the particulars of the partials of Δx and Δy depending on the distortion model being used. For the Acton–Duxbury model, Eq. (4.44) can be differentiated to yield

$$\begin{aligned} \frac{\partial x'}{\partial q} &= \left[1 - \alpha_1 xy/r + \alpha_2(r^2 + 2x^2) - 3\alpha_3 xyr + \alpha_4(r^4 + 4x^2y^2) + \alpha_5 y + 2\alpha_6 x\right]\frac{\partial x}{\partial q} \\ &\quad + \left[-\alpha_1(r + y^2/r) + 2\alpha_2 xy - \alpha_3(r^3 + 3y^2 r) + 4\alpha_4 xyr^2 + \alpha_5 x\right]\frac{\partial y}{\partial q} \quad \text{and} \end{aligned} \tag{6.9}$$

$$\frac{\partial y'}{\partial q} = \left[\alpha_1(r + x^2/r) + 2\alpha_2 xy + \alpha_3(r^3 + 3x^2 r) + 4\alpha_4 xyr^2 + \alpha_6 y\right] \frac{\partial x}{\partial q}$$
$$+ \left[1 + \alpha_1 xy/r + \alpha_2(r^2 + 2y^2) + 3\alpha_3 xyr + \alpha_4(r^4 + 4y^2r^2) + 2\alpha_5 y + \alpha_6 x\right] \frac{\partial y}{\partial q}. \tag{6.10}$$

For the OpenCV model, the equation to be differentiated is (4.45), and the resulting partials are

$$\frac{\partial x'}{\partial q} = \left[1 + k_1(r^2 + 2x^2) + k_2(r^4 + 4x^2r^2) + k_3(r^6 + 6x^2r^4) + 2p_1 y + 6p_2 x\right] \frac{\partial x}{\partial q}$$
$$+ \left[-\alpha_1(r + y^2/r) + 2\alpha_2 xy - \alpha_3(r^3 + 3y^2 r) + 4\alpha_4 xyr^2 + \alpha_5 x\right] \frac{\partial y}{\partial q} \text{ and} \tag{6.11}$$

$$\frac{\partial y'}{\partial q} = \left[\alpha_1(r + x^2/r) + 2\alpha_2 xy + \alpha_3(r^3 + 3x^2 r) + 4\alpha_4 xyr^2 + \alpha_6 y\right] \frac{\partial x}{\partial q}$$
$$+ \left[1 + \alpha_1 xy/r + \alpha_2(r^2 + 2y^2) + 3\alpha_3 xyr + \alpha_4(r^4 + 4y^2r^2) + 2\alpha_5 y + \alpha_6 x\right] \frac{\partial y}{\partial q}. \tag{6.12}$$

Finally, the partials are transformed using the matrix **K** to give the desired result, the partial derivatives of the observed sample and line coordinates with respect to the dynamic parameters. Using JPL's 2×3 matrix which includes a column for the product xy,

$$\begin{pmatrix} \partial s/\partial q \\ \partial l/\partial q \end{pmatrix} = \mathbf{K} \begin{pmatrix} \partial x'/\partial q \\ \partial y'/\partial q \\ x'\partial y'/\partial q + y'\partial x'/\partial q \end{pmatrix}. \tag{6.13}$$

6.2 Star Partials

Stars are so distant that their apparent positions depend only extremely weakly on the location of the camera within the Solar System. The closest star to the sun, Proxima Centauri, has a parallax of approximately $0''.786$, implying a distance of 4.24 light years or 4.02×10^{13} km. Furthermore—and more important for the purposes of optical navigation—the celestial coordinates of the stars are now known, thanks to the Gaia spacecraft, to an accuracy that exceeds the centerfinding capabilities of most optical navigation cameras. This happy circumstance implies that the coordinates of stars which appear in optical navigation pictures may be assumed known, and there is seldom a need to solve for them.

Nevertheless, occasionally a star's catalogued position will be in error, and it may be necessary to solve for the star's coordinates. One can differentiate Eq. (4.6) to obtain the partials of the star's position vector:

$$\frac{\partial \mathbf{r}(t)}{\partial \alpha} = \frac{1\ \text{pc}}{\pi}\begin{pmatrix} -\sin\alpha\cos\delta \\ \cos\alpha\cos\delta \\ 0 \end{pmatrix}; \tag{6.14}$$

$$\frac{\partial \mathbf{r}(t)}{\partial \delta} = \frac{1\ \text{pc}}{\pi}\begin{pmatrix} -\cos\alpha\sin\delta \\ -\sin\alpha\sin\delta \\ \cos\delta \end{pmatrix}; \tag{6.15}$$

$$\frac{\partial \mathbf{r}(t)}{\partial \pi} = \frac{1\ \text{pc}}{\pi^2}\begin{pmatrix} -\cos\alpha\cos\delta \\ -\sin\alpha\cos\delta \\ -\sin\delta \end{pmatrix}. \tag{6.16}$$

These partials play the role of $\partial \mathbf{T}/\partial q$ in the previous section, and they are transformed into sample and line coordinates using equations in the previous section, beginning with (6.2).

6.3 Optical Partials

The so-called "optical parameters" are those which affect only the transformation of **A** into sample and line coordinates. These include the camera pointing angles, which affect the matrix **C**, and the various parameters that describe the camera itself and affect the projection of **A** into sample and line. As such, changes in these parameters affect the spacecraft trajectory and the target bodies' ephemerides only indirectly, through correlations that may exist between the optical parameters and the dynamic ones.

6.3.1 Pointing Angles

Partial derivatives of the predicted image locations with respect to the pointing angles depend on how the matrix C was constructed. In the most general case, **C** is the product of several elementary rotation matrices, so that partial derivatives exist with respect to each of the angles in the formulation. Suppose that $\mathbf{C} = \mathbf{A}\ \mathbf{R}_i(\theta)\ \mathbf{B}$, where **A** and **B** are rotation matrices not depending on θ (one of both of these could be the identity matrix) and θ is the angle whose partial derivatives are requested. These derivatives may be obtained easily enough. If $i = 1$ (that is, the rotation in question is about the first axis), then

$$\frac{d\mathbf{R}_1(\theta)}{d\theta} = \begin{pmatrix} 0 & 0 & 0 \\ 0 & -\sin\theta & \cos\theta \\ 0 & -\cos\theta & -\sin\theta \end{pmatrix} \tag{6.17}$$

and

$$\frac{\partial \mathbf{C}}{\partial \theta} = \mathbf{A}\frac{d\mathbf{R}_1(\theta)}{d\theta}\mathbf{C}. \tag{6.18}$$

The partials for rotations about the y- and z-axes are similar.

A more interesting case would arise if one of the coordinate axes of the camera coordinate system be defined by a possibly variable direction in space, for instance, the direction from the camera to a target body for the optical axis, or the direction from the camera to the sun for the camera's M- or N-axes. These partials, however, are seldom required. In operations, no matter how $\mathbf{C}$ is constructed, a pointing solution is performed in terms of either right ascension, declination, and twist or (better) in terms of elementary rotations about the camera body axes. In neither case would the axes depend on any parameters other than the pointing angles themselves.

6.3.2 Focal Length

A change in focal length is the same as moving the back wall of the camera obscura toward or away from the pinhole on the other side of the room. This motion simply moves each image radially, either toward or away from the optical axis, by an amount proportional to the distance of the image from the optical axis. The partials are found by differentiating Eq. (4.40) with respect to f:

$$\frac{\partial(x, y)}{\partial f} = \frac{1}{P_3}\begin{pmatrix} P_1 \\ P_2 \end{pmatrix}.$$

6.3.3 Distortion Parameters

These are the α_i of the Acton–Duxbury model (4.44) or the k_i and p_i of the OpenCV model (4.45). These partials are therefore

$$\frac{\partial(x', y')}{\partial \alpha_i} = \begin{pmatrix} -yr & xr^2 & -yr^3 & xr^4 & xy & x^2 \\ xr & yr^2 & xr^3 & yr^4 & y^2 & xy \end{pmatrix} \tag{6.19}$$

for the Acton–Duxbury model. For the OpenCV model,

$$\frac{\partial(x', y')}{\partial k_i} = \begin{pmatrix} r^2 & r^4 & r^6 & 0 & 0 & 0 \\ 0 & 0 & 0 & r^2 & r^4 & r^6 \end{pmatrix} \tag{6.20}$$

and

$$\frac{\partial(x', y')}{\partial p_i} = \begin{pmatrix} 2xy & r^2 + 2x^2 \\ r^2 + 2y^2 & 2xy \end{pmatrix}. \tag{6.21}$$

6.3.4 Pixel Layout ("K Matrix") Parameters

These are found by differentiating Eq. (4.47) with respect to the four (or six) elements of the matrix **K** which transforms the distorted x' and y' coordinates of an image into pixels:

$$\frac{\partial(s,l)}{\partial K_x, K_{yx}, K_{xy}, K_y, K_{xxy}, K_{yxy}} = \begin{pmatrix} x' & 0 & y' & 0 & x'y' & 0 \\ 0 & x' & 0 & y' & 0 & x'y' \end{pmatrix}. \tag{6.22}$$

6.3.5 Optical Axis Pixel Coordinates

These are found trivially from Eq. (4.47) as well:

$$\frac{\partial(s,l)}{\partial(s_0,l_0)} = \begin{pmatrix} 1 & 0 \\ 0 & 1 \end{pmatrix}. \tag{6.23}$$

6.4 Constructing the List of Estimated Parameters

The previous sections have presented the various ways of calculating the partial derivatives of the pixel coordinates of an image with respect to the various parameters that can affect those calculations. The orbit determination team, in operations, must decide how best to handle the incoming optical navigation data. Recall that the parameter list can include

- dynamic parameters, which affect the trajectory of the spacecraft or the ephemeris of the observed bodies;
- radio parameters, such as range biases or quasar coordinates, which affect the handling of radio metric data;
- body size, shape, and orientation parameters, which affect the appearance of the body's image and therefore the predicted location of limb points, terminator points, or landmarks;
- landmark coordinates, which obviously affect the predicted location of landmark images; and
- camera parameters, including pointing angles, which affect the calculation of sample and line.

That the initial state of the spacecraft should be estimated is obvious. Inclusion of other dynamic parameters or radio parameters is at the discretion of the analyst, and a detailed discussion is outside the scope of this little book. As for the optical parameters, the usual practice has been to calibrate each camera using pictures of star clusters, and then to hold the camera parameters fixed thereafter unless there is obvious evidence of variability.

The one decision that bears some thought is whether an orbit solution should estimate (actually, *re*-estimate) the camera pointing. If there are more than two or three stars in each picture, the camera pointing can be considered known well enough, and further estimation is usually not necessary. If there are no stars in the picture—a common occurrence when the spacecraft is close to a body and landmarks are used—there is a significant correlation between the camera pointing and the spacecraft's position relative to the center of the body. For instance, one could move the spacecraft to the right and aim the camera to the left, and the boresight of the camera would point to the same spot on the body's surface. A picture taken under this perturbed scenario would not be identical to one taken under nominal conditions, because of different foreshortening in the horizontal direction, but the difference would not be immediately obvious. The standard practice at JPL for spacecraft orbiting small bodies (for instance, Dawn at Vesta or Ceres) has been to approach the problem iteratively. First the optical navigation analysts will assume that the spacecraft trajectory is known, and they will use the accumulated observations of landmarks to solve for the camera pointing for each picture, for the coordinates of the landmarks, and for the rotation parameters of the body. The revised information, particularly the landmark coordinates, is passed along to the larger navigation team, who produce another orbit solution for the spacecraft. After several iterations a consistent solution emerges.

It is advisable to employ a number of orbit determination scenarios, each using a different set of estimated parameters, a different set of weights for the various data types (in particular, optical navigation imagery relative to radio tracking data), or different assumptions about the *a priori* knowledge of the estimated parameters. Analysts will examine how each scenario evolves as more information is added and choose the one which they judge to be "best," i.e. the one most likely to be closest to the unknown and unknowable truth.

6.5 Camera Calibration

Although a camera's desired optical specifications are known before manufacture, and the characteristics of its detector likewise, the actual values of the camera parameters may be somewhat different. Similarly, the actual orientation of the camera on the spacecraft will not correspond exactly to the blueprints. It is therefore essential to calibrate each camera in flight.

A camera calibration has two purposes: to determine accurate values for the focal length, pixel layout, and optical distortion parameters; and to determine deviations of the actual or "as-flown" camera pointing from the camera attitude either as reported by the spacecraft or as commanded.

The necessary math has already been presented, in particular for the partial derivatives of the image locations with respect to the camera parameters and pointing angles. This section is devoted instead to guidelines for performing a successful calibration.

A typical calibration for a narrow-angle camera requires pictures of one or more star clusters. The ideal cluster is what amateur astronomers would call an "open" or "galactic" cluster—not a globular cluster—whose angular diameter is a good match to the field of view of the camera. The goal is to obtain star images everywhere in the field, and especially in the corners where distortion is most evident. It is common to command small spacecraft turns between pictures, so that the camera will usually have nine different pointings, arranged in a 3×3 grid, with the spacing between centers about 1/4 of the field of view. If the chosen cluster is significantly smaller than the field of view, the most important offsets will put the cluster in each corner of the field, because the corners lie farthest from the optical axis and are therefore most subject to distortions. Using more pictures and more pointings will obviously yield better results, but in practice the number of pictures and number of pointings will be subject to spacecraft or downlink constraints. Thus if nine pointings are not possible, a fallback would be five, with the centers forming a quincunx.

The calibration itself proceeds much like any other optical navigation image processing and analysis. One must identify the catalogued stars in each picture and determine their observed pixel coordinates. Typically a preliminary pointing solution is performed. Then, using the revised pointing, the predicted image locations are updated, new residuals formed, and new partial derivatives computed. All these are input to a filter, which estimates the desired camera parameter and simultaneously re-estimates the pointing angles. The process may need to be iterated, particularly if the as-flown focal length is significantly different from the nominal.

Note that not all of the camera parameters can be unambiguously determined. To first order, pictures of star clusters yield the *pixel scale*, or the angular size of a pixel (say, in arc seconds or microradians per pixel). By comparing differences in star coordinates to differences in the pixel coordinates of the star images, or by rotating the camera about its optical axis, one can tell that pixels are rectangles or rhombi as opposed to squares. The pixel scale s, however, is the ratio of the size of a pixel to the focal length of the camera, a pure number, but the camera parameters themselves carry dimensions. In the Acton–Duxbury model, the scale in the horizontal direction is $s_x = 1/(fK_x)$. It is clear that f and K_x cannot be uniquely determined, but rather it is their product that is observed. A common approach in calibration is to hold K_x fixed at the reciprocal of the manufacturer's value for the horizontal dimension of a pixel. Solving for f alone therefore accommodates observed changes to s_x. If s_y, the scale in the vertical direction of the picture, happens to be different from s_x, the remedy in the Acton–Duxbury model is to add K_y to the solution.

A similar argument applies if the sample and line axes are not perpendicular. Standard practice at JPL in this case has been to leave K_{xy} fixed at zero and solve for K_{yx} Thus the twist angle specifies the orientation of the sample axis of the camera, and K_{yx} then determines the orientation of the line axis.

Using the protocol outlined above, the camera parameters in the solution are typically these, in order of importance:

- camera pointing (changes to three Euler angles) for each picture;
- focal length;
- cubic radial distortion, represented in the Acton–Duxbury model by α_2 and in the OpenCV model primarily by k_1;
- "tip–tilt" terms, arising when the detector is not perpendicular to the optical axis, and captured by α_5 and α_6 in the Acton–Duxbury model and by p_1 and p_2 in the OpenCV model;
- pixel aspect ratio, K_y;
- nonorthogonality of the sample and line axes in the detector, K_{yx};
- fifth-order radial distortion, α_4 for Acton–Duxbury or k_2 and k_5 in OpenCV; and
- the pixel coordinates (s_0, l_0) of the optical axis.

There are two ways to proceed. One could begin with a minimal set of parameters (pointing angles and f) and examine the postfit residuals in a so-called "quiver plot" that displays each residual as a vector, positioned at the location of the star in the field of view. Systematic trends in the plot would indicate that additional terms are needed, in which case additional runs, solving for more parameters, are made and the process repeated until no discernible trends remain. The other approach is to solve for everything initially, examine not the residuals but the size of the distortion parameters compared to their sigmas, and delete those which are not statistically significant. In either approach one must use good engineering judgment. The foregoing assumes that obvious outliers have already been eliminated from the data set: misidentified double stars or bright "cosmic ray hits" adjacent to a star can corrupt the star's center and produce a bad residual. Occasionally, for whatever reason, a particular star will have an erroneous catalogued position.

A useful analytical tool for assessing the quality of a camera calibration is the so-called "quiver plot." The axes of the plot are the camera's sample and line axes, and the residual for each star image is plotted as a vector emanating from the observed location of that image. These vectors are usually scaled up by a factor of 100–1000 to make them more easily visible. If the calibration is good, the plots will look random, with no systematic trends anywhere in the field. The presence of trends indicates that at least one more term needs to be added to the camera model.

For narrow-angle cameras it is usually safe to place (s_0, l_0) at the center of the picture, but if a wide-angle camera has significant radial distortion it is important to use the correct location of the optical axis. Forcing it to be at the center of the field can yield swirl patterns in the postfit residuals, and these go away once one adds (s_0, l_0) to the calibration. The navigation cameras on OSIRIS-REx needed this special treatment.

Once the camera parameters have been found, the final step in the calibration is to compare the *a posteriori* camera pointing to the commanded pointing, picture by picture, in order to obtain an average offset. A significant offset can be accommodated by updating the orientation of the camera in the spacecraft body coordinate system.

6.5.1 Time-Varying Camera Parameters

One may encounter a camera whose focal length changes with temperature. As the camera body expands or contracts, the distance from the optics to the detector will change, and unless all the bundles of light are parallel as they impinge on the detector, the picture will appear larger or smaller—in effect, the focal length has changed.

Similarly, some cameras on spacecraft exploring small bodies will have an adjustable focus, so that pictures of the surface taken at close distances will be in focus. The effective focal length for these cameras can be a function of the focus setting.

To compensate for these effects requires (i) sets of calibration pictures taken under differing conditions and analyzed separately and (ii) some method of compensating for the effect in subsequent navigation analysis. Part (i), although time-consuming, is straightforward, especially if sunlight can be used to warm the camera under controlled conditions. For part (ii), one workable approach is to correct the measured location of each image, moving it radially either toward or away from the optical axis by an amount linearly proportional to both the image's distance from the axis and to the change in focal length relative to some standard value. These techniques were used on some of the OSIRIS-REx cameras with acceptable results.

Appendix A

The Overlapping Plate Method

The "overlapping plate method," originally developed by Eichhorn (1960), exploits the fact that a star can have only one position at a time and thereby enforces constraints that can vastly improve the quality of photographic astrometry. Owen, Synnott and Null (1999) applied this method to astrometry for Solar System bodies.

A.1 Development of Fundamental Catalogs

To appreciate the benefits of the overlapping plate method, one must first know how photographic astrometry was developed and applied to the creation of star catalogs. The first modern catalogs were compiled by eye, using *transit circles*, telescopes which are constrained to lie in the plane of the meridian. The time at which a star crossed the observer's meridian provided the star's right ascension, given that the telescope's longitude was known. The altitude of the star, measured with microscopes looking at a giant protractor, provided the star's declination, given that the telescope's latitude was known. The telescope's latitude could be determined by observing circumpolar stars at "lower culmination," when they crossed the meridian moving west to east below the pole. The telescope's longitude could be determined only by observing moving objects—the sun, planets, and particularly the moon—as their positions as a function of time were presumably known. This arduous process culminated in catalogs containing positions and eventually proper motions for some thousands of stars. Subsequent analysis, using such catalogs from observatories around the world, produced a series of "fundamental catalogs" whose coordinates implicitly defined the orientation of the celestial x-, y-, and z-axes.

Spacecraft Optical Navigation, First Edition. William M. Owen Jr.

A.2 The Astrographic Catalog

When photographic plates became sensitive enough that astrophotography became feasible, in the last two decades of the nineteenth century, astronomers quickly realized that a much better catalog could be derived from pictures of the sky. Some 20 observatories, spanning latitudes from +60° (Helsinki) and +51° (Greenwich) to −34° (Cape Town) and −47° (Melbourne), were assigned bands of declination to be photographed and measured. The "Astrographic Catalog" was one of the first successful international scientific collaborations, as it ultimately produced data for over four million stars.

The sky was typically photographed using a "corner-in-center" overlap pattern: the center of one plate would appear near a corner of four other, overlapping plates. This pattern meant that any particular star would typically appear on four plates. Standard practice at the time—indeed, through the creation of the Yale Zone catalogs in the 1930s—was to reduce each plate separately, using a small set of reference stars (which had been observed with transit circles) to determine the mapping from the measured (x, y) coordinates of star images into right ascension and declination. As each star would be observed four times, the four independent solutions for its position would simply be averaged.

At least this was the intention. Most observatories wound up publishing only lists of (x, y) coordinates of star images, a decision that would prove beneficial.

A.3 Development of the Overlapping Plate Method

Eichhorn's (1960) breakthrough was the realization that *a star has only one position at a time*, not four. Furthermore, if a star does appear on four photographic plates, one could use all the star positions from all four plates to produce one position that would likely be even better than the average one. Not only would there be more reference stars in the data reduction, but also the uncatalogued stars can serve as tie points between plates. These extra constraints—a star has only one position at a time—enable one to solve for more distortion terms than would otherwise be possible. The result is a catalog that is not only more precise but presumably more accurate, as it should be less prone to systematic error.

Owen, Synnott and Null (1999) applied the overlapping plate method to astrometry for Solar System objects, and they began an observing program that would continue for 20 years, supporting Galileo, Cassini, Stardust, Deep Impact, and New Horizons, and also enabling hundreds of successful observations of occultations of stars by asteroids. The observing program was designed specifically for the overlapping plate method. Each object to be observed would be imaged at least twice, with the target offset from the center of the field in a different direction each time.

The size and direction of the offsets would be chosen in order to get the most reference stars into the reduction while keeping the offsets below 20% or so of the field.

A.4 Application to Groundbased Astrometry

To reduce the pictures for one object required several steps. First, an image processing program provided the sample and line coordinates of the center of each measurable starlike image in each picture. Next, the commanded pointing was used to predict which stars would appear in or near the field of view in each picture. A pattern-matching algorithm located the reference stars and the target in each picture and produced as well an *ad hoc* catalog of all images it could not identify. It could recognize multiple images of the same uncatalogued stars, as these help constrain the overlap.

The data reduction used Bierman's (1977) U–D factorization method. The first two parameters in the estimate are changes to the original values of the right ascension and declination (actually $\Delta\alpha\ \cos\delta$ and $\Delta\delta$). Next follow the "bias parameters": telescope pointing (three parameters, including twist, for each picture), the telescope focal length, and various higher-order distortion parameters as necessary. These distortions are modeled using products of Legendre polynomials,

$$\Delta s = \sum a_{ij} P_i(x) P_j(y) \text{ and}$$
$$\Delta l = \sum b_{ij} P_i(x) P_j(y),$$

as Legendre polynomials are orthogonal over the interval $(-1, +1)$. The variables x and y above are the sample and line coordinates, normalized to $(-1, +1)$. Each distortion term may be applied on a picture-by-picture basis or for the entire set.

Bierman's method allows for "stochastic parameters," originally intended to mean those whose values can change randomly with time but whose history is knowable if the data are sensitive enough. For astrometry, changes to right ascension and declination—for the asteroid, for the reference stars, and for the uncatalogued stars—are treated all together as stochastic parameters. When data for a new object are encountered, the filter writes the current state to a scratch file, zeroes out the corrections to the position, and resets the first two rows and columns of the covariance to their initial values: for catalogued stars, the variances in their positions; for other stars and for the target objects, a very high (effectively infinite) value.

After the last data point has been ingested, the filter contains the final values for all of the bias parameters and the final solution for the position of the last star in the list. All of the other positions, which had been saved during the course of filtering, are based on only those measurements that had been presented to the filter at that time. A "smoothing" process then recovers what those positions would have been, with benefit of all the data. Smoothing works backwards, beginning with the penultimate star. It uses the *changes* in the solution for the bias parameters from what they had been for each star to their final value. The saved first two rows of the covariance matrix give in essence the partial derivatives of each star's coordinates with respect to the other parameters. In this way all of the star data are brought to bear on the coordinates of all of the stars.

The coordinates for the target(s) of interest are introduced to the filter last of all, and so they are unaffected by the smoothing process. The real purpose of smoothing is to generate reliable postfit residuals for the stars.

Glossary

aberration, any effect in a camera which causes a picture to differ systematically from the ideal picture

active pixel sensor, a detector (usually CMOS) whose pixels are individually addressable and can be read or flushed in any order

albedo, the ratio of emitted light from a surface to incident light on that surface

altitude, the angle that a celestial body makes with the horizon

angle of emission, the angle between light emitted from a surface and the normal to that surface

angle of incidence, the angle between light impinging on a surface (usually from the sun) and the normal to that surface

apparent position, the direction from which light appears to come, after accounting for the effects of light time and stellar aberration

astigmatism, the third-order optical aberration which causes out-of-focus images to appear elliptical rather than circular

barrel distortion, the third-order optical aberration which causes squares to appear rounded, because images far from the optical axis are moved inwards; cf. pincushion distortion

bias, a constant negative voltage added to the output of an electronic detector to ensure that the final voltage will remain negative even in the presence of read noise

bias pictures, digital images taken without opening the shutter, to establish the average “zero point” for each pixel

camera, a physical device which captures a picture of the scene in front of it

central epoch, the time at which a star’s position is best known, equal to the weighted average of the times of all the observations that were used in determining the position and proper motion

charge-coupled device (CCD), a light-sensitive silicon chip divided by high-voltage “gates” into rectangular regions which contain photoelectrons liberated during an exposure; the voltage on the gates is varied systematically

Spacecraft Optical Navigation, First Edition. William M. Owen Jr.

to shift the contents of each pixel into a readout amplifier, thereby producing a digital image

coma, the third-order optical aberration which causes light entering farther from the center of the aperture to impinge on the detector farther away from the optical axis

complementary metal-oxide semiconductor (CMOS), a light-sensitive silicon chip (cf. CCD) in which each pixel contains its own readout port, enabling pixels to be read in any order without requiring the electrons to be moved through other pixels for readout

conduction band, in a semiconductor, an energy level in which electrons can move freely rather than being bound to an atom

correlation peak, the maximum value when two functions (or two data sets) are correlated against each other, or the coordinates at which this maximum occurs

correlation score, the result of correlating two functions (or two data sets) at a specific offset

curvature of field, the third-order optical aberration which causes best focus to lie not on a plane but on the surface of a sphere; when present, a planar detector cannot produce in-focus images everywhere in the picture

cusps, the points at which the limb and terminator of a body may intersect

dark limb, the unlit portion of the limb of a body; cf. lit limb

data number (DN), the digitized output of the signal in a pixel; more generally, the digitized output from any sensor on a spacecraft

detector, in the context of optical navigation, a device placed in the focal plane of a camera whose function is to record incident light as a function of position in the focal plane

dichotomy, in astronomy, the geometry at which one half of the visible surface of an object is sunlit

distortion, the third-order optical aberration in which light impinges on a detector at a distance from the optical axis other than that predicted by the gnomonic projection

elevation, the distance of a point on the surface of a body above or below the reference model (for the earth, relative to sea level); also, one of the two angles used to characterize the pointing of some cameras

ellipsoid, a body characterized by the equation $(x/a)^2 + (y/b)^2 + (z/c)^2 = 1$, with a, b, and c not necessarily equal

exposure, the length of time at which the detector in a camera is allowed to have light impinge upon it; also a synonym for the resulting picture

field of view, the volume (typically a rectangular pyramid) which projects onto the detector of a camera

fixed pattern noise, a constant unequal background present in every picture, usually caused by manufacturing irregularities in the detector
focal plane, in a camera, the plane at which images are in best focus; also, incorrectly, the (assumed) plane of the detector
forward scattering, the tendency of light impinging on the surface of a body to be re-emitted with little change in its direction
gain, in a digital camera, the number of electrons per DN
gates, usually rectangular epitaxial layers on the surface of a CCD or CMOS detector, whose voltage is controllable, used to confine electrons within a pixel during an exposure and (in a CCD) to shift electrons during readout
geometric optics, the branch of optics which deals with formation of images without regard to the photon nature of light
geometric position, the vector (or spherical coordinates) from a camera to an object's physical position at the time of the exposure
grazing incidence, the geometry at which incoming light is nearly parallel to a surface
hole, in a semiconductor, the positive charge left behind when an electron in a silicon atom obtains enough energy to leave its atom for the conduction band
hot pixels, those pixels in an electronic detector which accumulate charge at an abnormally high rate even in the absence of incident light
image, a pattern of light on a detector which represents the appearance of some object in the field of view of the camera
image processing, for optical navigation purposes, the extraction of optical navigation observables (the sample and line coordinates of images within a picture) from a digital picture
landmark, a uniquely identifiable feature or area on the surface of a body
landmark map, the digital representation of a landmark, typically containing a square array of relative heights and relative albedos
limb, the set of points on a body such that a line drawn from the observer to a limb point is tangent to the body's surface
lit limb, the set of limb points which are also sunlit; cf. dark limb
magnitude, the integrated brightness of a star or other celestial body, expressed in an arbitrary logarithmic scale such that a difference of 5 magnitudes corresponds to a factor of 100 in brightness; originally defined such that Vega was magnitude 0.0 in all passbands, but now defined implicitly by a set of standard stars
marginal distribution, in probability theory, a probability distribution that results from integrating a multivariate distribution function of some but not all of the variables; for optical navigation purposes, summing the brightness in a picture (or a subset of a picture) in either the sample or line directions, leaving a one-dimensional array of observed brightnesses

master bias, a picture constructed from a set of bias pictures, intended to provide a better representation of the bias
normal incidence, the geometry such that incoming sunlight is perpendicular to a body's surface
object, any physical thing; in the context of optical navigation or photography, a physical thing in the field of view of a camera
observable (noun), in spacecraft navigation, any quantity which is amenable to observation or measurement and whose value is sensitive to parameters which model the motion of the spacecraft or the motion or physical characteristics of the bodies being observed
opposition surge, the tendency of light near normal incidence to be scattered preferentially toward the normal direction
optical navigation, the branch of spacecraft navigation concerned with the planning, prediction, and processing of pictures taken by cameras on a spacecraft
parallax, the change in apparent position of an object arising from the change in position of the observer; in particular, a star's *annual parallax* results from the motion of the earth around the sun
parent, the entity to which an object's position and velocity are referred
phase angle, the angle, usually at the surface of a body, between the direction to the sun and the direction to an observer
photocenter, for a body whose angular diameter is so small that it appears starlike, the average direction of the light from that body as received by an external observer
photoelectrons, in a CCD or CMOS detector, electrons that have received enough energy from an incident photon to have been moved into the conduction band
picture, a two-dimensional representation of a part of the three-dimensional universe
pincushion distortion, the third-order optical aberration which causes the sides of squares to appear concave, because images far from the optical axis are moved outwards; cf. barrel distortion
pixel, short for "picture element," one of many light-sensitive areas of a detector that produces a DN value indicative of the amount of light incident on it during an exposure
pixel scale, the physical size of a pixel, often measured in microns
point-spread function, a mathematical model which describes how the image of a point source actually appears within a picture
port, in a CCD, the electronics by which the amount of accumulated charge in each pixel is made available
principal axes, the three mutually perpendicular axes of a rigid body such that the inertia tensor is diagonal

principal point, the intersection of the optical axis of a camera with the image plane
proper motion, the angular rate of a star, caused by its motion relative to the solar system
pseudo-rectangular plate, a patch of a body's surface defined by four points, with interior points determined using bilinear interpolation
read noise, the inability of an electronic detector to return the exact number of photoelectrons in a pixel, independent of the number of them; usually assumed to follow a Gaussian distribution
readout register, in a CCD, a one-dimensional array of pixel-like regions into which each row in the picture is temporarily stored prior to readout; the electrons are then moved, pixel by pixel, from the readout register into the port
real image, in optics, the locus of points at which all rays emanating from any point in a scene converge
resolution, for a camera, the angle subtended by one pixel; a body is usually said to be *resolved* if the diameter of its image exceeds two pixels
retarded time, the time at which light leaves an object being imaged
rotation matrix, a 3×3 orthonormal matrix which, when multiplying a vector, transforms its rectangular coordinates from one coordinate system to another
scan angle, the angle within a picture which a scan line makes with the horizontal axis of the picture
scan line, a line (actually a ray) within a picture, emanating from the assumed center of the image of a body at a specified angle
scene, the set of objects that are imaged by a camera
space velocity, the three-dimensional velocity of a star, usually relative to the barycenter of the Solar System
specular reflection, from the Latin *speculum* "mirror," the emission of reflected light such that (i) the angle of emission equals the angle of incidence and (ii) the incoming and outgoing rays are coplanar
spherical aberration, the third-order optical aberration such that parallel rays of light entering at different distances from the center of the aperture are focused at different distances from the aperture, and a point source is never truly "in focus"
spheroid, a body characterized by the equation $(x/a)^2 + (y/a)^2 + (z/c)^2 = 1$, with a and c not necessarily equal; a spheroid is *oblate* if $c < a$, *prolate* if $c > a$
stellar aberration, the optical phenomenon which causes incoming light to appear to be shifted toward the direction of motion of the camera
stereophotoclinometry, a technique for modeling the surface of a body, combining photoclinometry ("shape through shading") and stereo imaging
surface normal, of a point on a body, the vector which is perpendicular to the plane which is tangent to the surface at that point

target, any object or portion of an object which is intended to be imaged
terminator, the set of points on a body which divides the sunlit surface from the dark (unlit) surface
triangular plate, a patch of a body's surface defined by three points, with interior points lying on the plane determined by these three points
triaxial ellipsoid, the set of points which satisfy an equation of the form $(x/a)^2 + (y/b)^2 + (z/c)^2 = 1$ with a, b, and c unequal
true position, the direction from which light from an object reaches an observer, with light time taken into account but with stellar aberration ignored
vertex (plural *vertices*), in three dimensions, a point at which three or more planes intersect

References

Archinal, B. *et al.* 2010. *Celest. Mech. Dyn. Astron.* **109**, 101, doi: 10.1007/s10569-010-9320-4.

Bhaskaran, S., Riedel, J. E., and Synnott, S. P. 1995. "Demonstration of Autonomous Orbit Determination Around Small Bodies," paper AAS 95–387, AAS/AIAA Astrodynamics Specialist Conference, Halifax, Nova Scotia.

Bhaskaran, S., Riedel, J. E., and Synnott, S. P. 1997. "Autonomous Nucleus Tracking for Comet/Asteroid Encounters: The STARDUST Example," paper AAS 97–628, AAS/AIAA Astrodynamics Specialist Conference, Sun Valley, Idaho.

Bhaskaran, S. *et al.* 1998. "Orbit Determination Performance Evaluation of the Deep Space 1 Autonomous Navigation System," paper AAS 98–193, AAS/AIAA Spaceflight Mechanics Meeting, Monterey, California.

Bhaskaran, S. *et al.* 2000. "The Deep Space 1 Autonomous Navigation System: A Post-flight Analysis," paper AIAA 2000–3935, AIAA/AAS Astrodynamics Specialist Conference, Denver.

Bhaskaran, S. *et al.* 2002. "Navigation of the Deep Space 1 Spacecraft at Borrelly," paper AIAA 2002–4815, AAS/AIAA Astrodynamics Specialist Conference, Monterey, California.

Bhaskaran, S. *et al.* 2004. "Optical Navigation for the STARDUST Wild 2 Encounter," 18th International Symposium on Spaceflight Dynamics, Munich.

Bradski, G. 2000. "The OpenCV Library," *Dr. Dobb's Journal of Software Tools.*

Breckenridge, W. G., and Acton, C. H. Jr. 1972. "A Detailed Analysis of Mariner Nine TV Navigation Data," paper 72–866, AIAA Guidance and Control Conference, Stanford.

Campbell, J. K. *et al.* 1980. "Voyager 1 and Voyager 2 Jupiter Encounter Orbit Determination," paper AIAA–80–0241, AIAA 18th Aerospace Sciences Meeting, Pasadena.

Spacecraft Optical Navigation, First Edition. William M. Owen Jr.

Campbell, J. K. *et al.* 1982. "Voyager 1 and Voyager 2 Saturn Encounter Orbit Determination," paper AIAA–82–0419, AIAA 20th Aerospace Sciences Meeting, Orlando.

Cheng, A. F. *et al.* 2008. *Space Sci. Rev.* **140**, 89, doi: 10.1007/s11214-007-9271-6.

Duxbury, T. C. 1970. "Navigation Data from Mariner Mars 1969 TV Pictures," *Navigation* **17**, No. 3, 219–225.

Duxbury, T. C., and Acton, C. H. Jr. 1972. "On-Board Optical Navigation Data from Mariner 71," *Navigation* **19**, No. 4, 295–307.

Duxbury, T. C., and Breckenridge, W. G. 1970. "Mariner Mars 1969 Optical Approach Navigation," AIAA paper 70–70, AIAA 8th Aerospace Sciences Meeting, New York.

European Space Agency 1997. *The HIPPARCOS and Tycho Catalogues*, ESA SP–1200.

European Space Agency 1997. *The Hipparcos and Tycho Catalogues*, document SP–1200.

Gaia Collaboration 2018. "Gaia Data Release 2. Summary of the Contents and Survey Properties," *Astron. Astrophys.* **616**, 1.

Gaskell, R. W. 2011. "Optical Navigation Near Small Bodies," AAS paper 2011–220, 21st AAS/AIAA Space Flight Mechanics Meeting, New Orleans.

Gaskell, R. W., Barnouin-Jha, O. S., and Scheeres, D. J. 2007. "Modeling Eros with Stereophotoclinometry," 38th Lunar and Planetary Science Conference, League City, Texas, p. 1333.

Gillam, S. D. *et al.* 2007. "Optical Navigation for the Cassini/Huygens Mission," AAS paper 07–252, AAS/AIAA Astrodynamics Specialists Conference, Mackinac Island, Michigan.

Hapke, B. 2002. *Icarus* **157**, 523, doi: 10.1006/icar.2002.6853.

Howell, S. B. 2006. *Handbook of CCD Astronomy*, 2nd edition (Cambridge: Cambridge University Press).

Itseez, Inc. 2015. *Open Source Computer Vision Library*, https://github.com/itseez/opencv.

Janesick, J. R. *et al.* 1990. *Proc. SPIE* **1242**, 223, doi: 10.1117/12.19452.

Jenkins, F. A., and White, H. E. 1976. *Fundamentals of Optics*, 4th edition (New York: McGraw–Hill).

Jerath, N. 1978. *Interplanetary Approach Optical Navigation with Applications*, JPL Publication 78–40.

Kizner, W. 1961. "A Method of Describing Miss Distances for Lunar and Interplanetary Trajectories", *Planet. Space Sci.* **7**, 125–131.

Mastrodemos, N. *et al.* 2006. "Autonomous Navigation for Deep Impact," AAS paper 06–177, AAS/AIAA 16th Space Flight Mechanics Conference, Tampa.

Mastrodemos, N. *et al.* 2012. "Optical Navigation for the Dawn Mission at Vesta," 23rd International Symposium on Space Flight Dynamics, Pasadena.

Minnaert, M. 1941. *Astrophys. J.* **93**, 403, doi: 10.1086/144279.

Morabito, L. A. *et al.* 1979. "Discovery of Currently Active Extraterrestrial Volcanism," *Science* **204**, 972, doi: 10.1126/science.204.4396.972.

Owen, W. M. Jr., Synnott, S. P., and Null, G. W. 1998. "High-Accuracy Asteroid Astrometry from Table Mountain Observatory," in *Modern Astrometry and Astrodynamics*, R. Dvorak, H. F. Haupt, and K. Wodnar, eds. (Verlag der Österreichischen Akademie der Wissenschaften, Vienna), 89–102.

Owen, W. M. Jr., Synnott, S. P., and Null, G. W. 1999. "High-Accuracy Asteroid Astrometry From Table Mountain Observatory," in *Modern Astrometry and Astrodynamics, Proceedings of the International Conference honoring Heinrich Eichhorn*, Held at Vienna Observatory, Austria, 25–26 May, 1998, R. Dvorak, H. F. Haupt, and K. Wodnar, eds.

Owen, W. M. Jr. *et al.* 2001. "NEAR Optical Navigation at Eros," paper 01–376, AAS/AIAA Astrodynamics Specialists Conference, Quebec City, Quebec.

Owen, W. M. Jr. *et al.* 2006. "Optical Navigation for Deep Impact," AAS paper 06–176, AAS/AIAA 16th Space Flight Mechanics Conference, Tampa.

Owen, W. M. Jr., Duxbury, T. C., Acton, C. H. Jr., Synnott, S. P., Riedel, J. E., and Bhaskaran, S. 2008. "A Brief History of Optical Navigation at JPL," paper AAS 08–053, 31st Annual AAS Guidance and Control Conference, Breckenridge, Colorado.

Owen, W. M. Jr., Weaver, H. A., and Cheng, A. 2019. *Publ. Astron. Soc. Pac.* **132**, 14504, doi: 10.1088/1538-3873/ab583f.

Rayman, M. D. *et al.* 2000. "Results from the Deep Space 1 Technology Validation Mission," *Acta Astronaut.* **47**, 475–487.

Riedel, J. E., Owen, W. M. Jr., Stuve, J. A., Synnott, S. P., and Vaughan, R. M. 1990. "Optical Navigation During the Voyager Neptune Encounter," paper 90–2877, AIAA/AAS Astrodynamics Conference, Portland, Oregon.

Ruyten, W. 1999. *Opt. Lett.* **24**, 878, doi: 10.1364/OL.24.000878.

Smithsonian Astrophysical Observatory 1966. *Smithsonian Astrophysical Observatory Star Catalog.*

Synnott, S. P. 1980. "1979J2—Discovery of a Previously unknown Jovian Satellite," *Science* **210**, 786–788, doi: 10.1126/science.210.4471.786.

Synnott, S. P. 1981. "1979J3—Discovery of a Previously unknown Satellite of Jupiter," *Science* **212**, 1392, doi: 10.1126/science.212.4501.1392.

Synnott, S. P. *et al.* 1986. "Interplanetary Optical Navigation: Voyager Uranus Encounter," paper 86-3113, AIAA/AAS Astrodynamics Conference, Williamsburg, Virginia.

Van Allen, R. E., Cesarone, R. J., and Gray, D. L. 1982. "Voyager Uranus–Neptune Navigation Planning Document," JPL 618–125 (internal document).

Vaughan, R. M. *et al.* 1992. "Optical Navigation for the Galileo Gaspra Encounter," AIAA paper 92–4522, AIAA/AAS Astrodynamics Conference, Hilton Head, South Carolina.

von Lommel, E. 1887. *Sitzber Acad. Wissensch. München* **17**, 95.

Williams, B. *et al.* 2018. *Space Sci. Rev.* **214** (4), 69, doi: 10.1007/s11214-018-0501-x.

Woodhouse, C. 2017. *The Astrophotography Manual: A Practical and Scientific Approach to Deep Space Imaging*, 2nd edition (Waltham, Massachusetts: Focal Press).

Zacharias, N. *et al.* 2004. "The Second US Naval Observatory CCD Astrograph Catalog (UCAC2)," *Astron. J.* **127**, 3043–3059, doi: 10.1086/386353.

Further Reading

Acton, C. H. Jr. 1972. "Processing Onboard Optical Data for Planetary Approach Navigation," *J. Spacecr. Rockets* **9**, No. 10, 746–750, doi: 10.2514/3.61796.

Eichhorn, H. K. 1990. *Astron. Nachr.* **285**, 233.

Hapke, B. 1981. *J. Geophys. Res.* **86**, 3039, doi: 10.1029/JB086iB04p03039.

Hapke, B. 1984. *Icarus* **59**, 41, doi: 10.1016/0019-1035(84)90054-X.

Hapke, B. 1986. *Icarus* **67**, 264, doi: 10.1016/0019-1035(86)90108-9.

Hapke, B. 1990. *Icarus* **88**, 407, doi: 10.1016/0019-1035(90)90091-M.

Hapke, B. 1999. *J. Quant. Spectrosc. Radiat. Transfer* **61**, 565, doi: 10.1016/S0022-4073(98)00042-9.

Lambert, J. H. 1760. *Photometria, sive de Mensura et gradibus luminis, colorum et umbrae* (Augsburg: Christoph Peter Detleffsen).

Owen,W. M. Jr. 2011. "Methods of Optical Navigation," AAS paper 2011–215, 21st AAS/AIAA Space Flight Mechanics Meeting, New Orleans.

Tabel, S., and Stechele,W. Proceedings of the 2017 International Conference on Computer Graphics and Digital Image Processing (Prague, 2017).

Index

Printed and bound by CPI Group (UK) Ltd, Croydon, CR0 4YY

17/09/2026

14964255-0008